Mathematical Modelling and Physical Applications of Magnetic Systems

Mathematical Modelling and Physical Applications of Magnetic Systems

Guest Editors

Roberto Zivieri
Giancarlo Consolo
Israa Medlej

MDPI

Basel • Beijing • Wuhan • Barcelona • Belgrade • Novi Sad • Cluj • Manchester

Guest Editors

Roberto Zivieri
Department of Mathematical and Computer Sciences, Physical Sciences and Earth Sciences
University of Messina
Messina
Italy

Giancarlo Consolo
Department of Mathematical and Computer Sciences, Physical Sciences and Earth Sciences
University of Messina
Messina
Italy

Israa Medlej
Shenzhen Institute for Quantum Science and Engineering
Southern University of Science and Technology
Shenzhen
China

Editorial Office

MDPI AG
Grosspeteranlage 5
4052 Basel, Switzerland

This is a reprint of the Special Issue, published open access by the journal *Magnetism* (ISSN 2673-8724), freely accessible at: https://www.mdpi.com/journal/magnetism/special_issues/MagneticSystem.

For citation purposes, cite each article independently as indicated on the article page online and as indicated below:

Lastname, A.A.; Lastname, B.B. Article Title. *Journal Name* **Year**, *Volume Number*, Page Range.

ISBN 978-3-7258-7074-5 (Hbk)
ISBN 978-3-7258-7075-2 (PDF)
https://doi.org/10.3390/books978-3-7258-7075-2

Contents

About the Editors

Roberto Zivieri

Roberto Zivieri is a theoretical condensed matter physicist. He received a Master's Degree in Medicine and Surgery and in Physics with honors, and a Ph.D. in Physics (grade excellent) from the University of Modena, Italy. His main research interests are phonon dynamics and scattering cross-section lattices, theoretical magnetism and microscopic magnetic models, magnonic crystals and spin excitations in periodic magnetic systems, scattering of light on magnetic systems and scattering cross-section, topological defects in condensed matter physics with special regard to 2D magnetic skyrmions, quantum magnetic models, classical and quantum statistical thermodynamics, the second law of thermodynamics, magnetic models applied to naval systems, and theoretical acoustics and applications in acoustics transducers. He is the author of around 200 scientific contributions in international and reputed journals. He serves on a respected Editorial Board. He is a Project Manager of scientific international projects. He is a member of the APS, ACS, Italian Society of Mathematical Physics, Italian Society of Physics, and Italian Society of Magnetism. He was the winner of the APS Award "Outstanding Referees 2016," equivalent to a journal APS Fellowship; of the Albert Nelson Lifetime Achievement Award; and is a Fellow of the International Association of Advanced Materials. He was Editor-in-Chief of the book series "*Theoretical and Computational Models in Condensed Matter Physics*" for Aracne Publishing House in Rome. His teaching has focused on solid state physics, the structure of matter, and the physics of critical phenomena.

Giancarlo Consolo

Giancarlo Consolo was born in Messina on June 24, 1980. He graduated summa cum laude in Electronic Engineering from the University of Messina in 2004 and earned a PhD in Advanced Technologies for Optoelectronics, Photonics and Electromagnetic Modeling in 2008. He became a permanent Researcher in Mathematical Physics in 2011 and has served as an Associate Professor of Mathematical Physics at the University of Messina since 2016. In 2009, he received the Young Researcher Award from the University of Messina.

He has led several competitive research projects funded by INdAM-GNFM and the Italian PRIN program. These include projects on mathematical models in biology, ecology, and fluid dynamics; emergent behaviors in hyperbolic reaction–diffusion systems; multiscale phenomena in continuum mechanics (PRIN 2017); ecosystem resilience under climate change (PRIN 2022 PNRR); and mathematical modeling for biofuel production (PRIN 2022 – MOMENTA). He was also a member (2010–2020) of the academic spin-off Goparallel s.r.l., established with the University of Salamanca.

Prof. Consolo has played major organizational roles in international conferences and workshops and has served as evaluator of national and international research projects since 2013. He is a member of INdAM-GNFM, the European Society for Mathematical and Theoretical Biology, and the Accademia Peloritana dei Pericolanti.

He serves on Editorial Boards of international journals and regularly reviews for leading journals in physics and applied mathematics.

His research focuses on reaction–diffusion and reaction–transport models in biology and ecology, micromagnetism and spintronics, and magnetoelastic phenomena. He has authored around 100 peer-reviewed papers and 4 book chapters, and has delivered numerous invited and contributed talks.

Israa Medlej

Israa Medlej is a former Postdoctoral Researcher in Advanced Spin-Wave Research: Experiments and Micromagnetic Modeling, Spintronics and Magnonics at the esteemed International Quantum Academy, Southern University of Science and Technology (SUSTech), Shenzhen, China. She holds a Ph.D. in Physics from the University of Messina, Italy, where she investigated magnetic skyrmion dynamics and THz oscillations in antiferromagnets. Her scientific activity includes micromagnetic simulations using MATLAB and other platforms to study magnetic skyrmions, as well as self-taught expertise in Mumax3 for modeling antiferromagnetic materials and topological spin textures like magnetic hopfions and torons. Her interdisciplinary research bridges theory and experiment, with a focus on spin-wave-driven phenomena for quantum computing. She has authored several peer-reviewed publications and received numerous awards, including High-Level Foreign Talents (Category A)—Southern University of Science and Technology, 2021, Shenzhen, China; Erasmus + International Credit Mobility Student [KA107]—University of Santiago de Compostela, 2019, Spain; Best Poster Award—Italian Magnetism Association Conference (MAGNET 2019), 2019, Messina, Italy; Excellence Ph.D. Dissertation—University of Messina and EDST, 2019, Messina, Italy; Best Oral Presentation—Research Day in Forbach, 2016, France; and Master 2 Internship Grant—University of Lorraine, 2016, Saint Avold, France. She actively contributes to the scientific community through conference presentations, manuscript reviews for journals such as *PRL*, *PRB*, and many other journals, and editorial roles.

magnetism MDPI

Editorial

Mathematical Modelling and Physical Applications of Magnetic Systems

Roberto Zivieri [1,*], Israa Medlej [2] and Giancarlo Consolo [3]

1 Department of Engineering, University of Messina, Contrada Di Dio, Vill. Sant' Agata, 98166 Messina, Italy
2 Shenzhen Institute for Quantum Science and Engineering, Southern University of Science and Technology, Shenzhen 518055, China; medlej.israa@gmail.com
3 Department of Mathematical and Computer Sciences, Physical Sciences and Earth Sciences, University of Messina, Contrada Di Dio, Vill. Sant' Agata, 98166 Messina, Italy; giancarlo.consolo@unime.it
* Correspondence: roberto.zivieri@unife.it

1. Introduction

Recently, nanoscale magnetic materials have attracted widespread attention due to their intriguing properties in both theoretical and experimental contexts. During the last two decades, the underlying physics of the complex magnetization dynamics in magnetic systems, such as ferromagnetic and antiferromagnetic films and confined magnetic systems of different shapes, including nanopillars and waveguides, has been widely studied via the formulation of sophisticated mathematical models, both in terms of classical and quantum descriptions. On the other hand, during the same period, intense experimental activity has successfully confirmed the existence several phenomena predicted by the theoretical models and revealed their effects. Concrete applications, such as magnetic memory, logic devices, diodes, transistors, microwave oscillators, and spintronic devices, etc., have been observed. Two main physical phenomena—spin transfer torque (STT) and spin-Hall effects (SHE)—paved the way for the fabrication of spintronic devices. The main goal of this Special Issue of *Magnetism*, "Mathematical Modelling and Physical Applications of Magnetic Systems," was to encourage world-leading scientists to present their latest exciting theoretical and experimental results at the nanoscale dimension, discussing the fundamental physics of these systems. The collection of articles highlights theoretical, experimental, and technological applications in the field of low-dimensional magnetic systems; the contributions are summarized in the following sections.

2. The Present Special Issue

In the SI "Mathematical Modelling and Physical Applications of Magnetic Systems" 9 papers were selected for publication following a strict review process. The main findings of these articles are described and discussed below.

However, before delving into the realm of the published articles, it is important to address three main inquiries: (1) What are the main arguments discussed, and why are they so important for magnetic systems? (2) Are these systems 1D, 2D, or 3D or all of them? (3) If they are also 3D, what are the advantages of studying them compared to low-dimensional 1D and 2D magnetic systems?

(1) What are the main arguments discussed and why are they so important for magnetic systems?

 https://doi.org/10.3390/magnetism6010002

In this Editorial, the main topics addressed are the mathematical modeling, deep phenomenological physics, and technological application of magnetic systems at different scales of dimensionality. Accurate descriptions of magnetization dynamics, spin–orbit coupling, nonlinear effects, and topological phenomena are gathered to tailor strong advanced theoretical approaches, such as the ones based on the extension of the Landau–Lifshitz–Gilbert (LLG) equation.

These arguments are crucial for

- A better understanding of magnetism at low dimensions, taking into account that theory and experiments need to be correlated.
- A deeper comprehension of physical phenomena, such as spin transport, spin torque effects (STT, SHE), and emergent electrodynamics, which are crucial for tailoring spintronic devices.
- Wider use of technological applications ranging from magnetic memory, logic devices, to neuromorphic computing with special regard to quantum technology based on 3D solitons like hopfions.

Topological spin-textures, magnetoelastic effects, and thermal–magnetic coupling are promising candidates for the development of next-generation technological applications correlated with magnetic materials.

(2) Are the magnetic topological systems studied 1D, 2D, or 3D—or all of them?

This collection investigates 1D, 2D, and 3D magnetic topological systems.

- 1D magnetic topological systems consist of domain walls and nanowires and were studied in the framework of spin transport and torque.
- 2D magnetic topological systems are mainly represented by magnetic skyrmions hosted in thin films, multilayers, and planar structures characterized by collective excitations such as spin waves.
- 3D magnetic topological magnetic systems are mainly represented by magnetic hopfions and torons and reveal a novel way to provide topologically protected knots in three dimensions.

One crucial consideration is the search for a novel approximation based on the inclusion of spin–orbit and topological torques in the LLG equation, which leads to a unified theory for 1D, 2D and 3D magnetic systems.

(3) If they are also 3D magnetic systems, what are the advantages of studying them compared to the low-dimensional 1D and 2D magnetic systems?

- Topological richness: 3D systems have a topologically complex spin-texture, such as a magnetic hopfion, characterized by the Hopf invariant, where magnetic field lines are linked. This phenomenon cannot exist in 1D and 2D.
- Enhanced stability: greater stability is one of the main factors that makes 3D systems more robust in relation to thermal fluctuations and defects, due to their volumetric topology.
- Improved information density: through the 3D profile, a high storage capacity can be achieved.
- New functionalities: 3D topological Hall transport and volumetric spin-wave modes propagation can be obtained through the coupling between magnetic, electric, and optical fields, paving the way for novel control mechanisms.

 https://doi.org/10.3390/magnetism6010002

3D magnetism is harnessed for technological applications (spintronics, magnonics, neuromorphic architectures, and quantum devices), suppressing the limitation of a planar geometry.

The topics can be subdivided into three main subjects:

1. **Advanced Magnetic Phenomena and Theoretical Insights**

Included Articles

- Magnetic Hopfions: A Review (Guslienko, 2024).
- A Study on the Effect of Plastic Strain on Magnetic Phenomenology and Microstructure (Hasičić et al., 2025).
- Can We Still Find an Ideal Memristor? (Wang, 2024).

2. **Magnetic Materials, Devices, and Applications**

Included Articles

- Loss Mitigation in Self-Biased Microstrip Circulators (Kong et al., 2023).
- Practical Study of Mixed-Core High Frequency Power Transformer (Paul, 2022).
- Magneto Elasticity Modeling for Stress Sensors (Diguet et al., 2022).

3. **Magnetic Modelling and Computational Methods**

Included Articles

- Vector-Based Magnetic Circuit Modelling of Induction Motors (Kidd, 2022).
- Numerically Stable and Computationally Efficient Expression for the Magnetic Field of a Current Loop (Ortner et al., 2022).
- The Role of Blood Perfusion in the Thermal Interaction Between Magnetic Nanoparticles and Cancerous Tumors (Maniotis et al., 2025).

The review article by K. Guslienko, ***Magnetic Hopfions: A Review***, deals with magnetic hopfions, which are the 3D topological counterpart of well-known 2D topological spin textures, such as magnetic skyrmions. Magnetic hopfions can be classified as field magnetic solitons because they exhibit stable and localized spin configurations that maintain their structure under continuous magnetic field deformation. Their creation can be numerically reproduced by means of micromagnetic simulations using software such as Mumax3 [1] or OOMMF [2]. On the other hand, their theoretical description is based on the LLG equation, which governs the magnetization motion. The 3D topological charge, also known as the Hopf index, can be extracted through the lens of emergent magnetic vector potential and magnetic field. Magnetic hopfions are physically explained using classical field theory and with toroidal coordinates which become crucial for properly describing them. Their stability and dynamics, especially in ferromagnetic films and dots, are discussed.

In M. Hasičić et al.'s article, "***A Study on the Effect of Plastic Strain on Magnetic Phenomenology and Microstructure***", it is revealed that magnetic properties are affected by plastic strain, correlating macroscopic behavior with microscopic mechanisms. The effect of the strain on the magnetization depends on the anisotropy profile. Based on micromagnetic simulations of hysteresis loops for different magnetic anisotropies, the different equilibrium configurations are discussed as a function of anisotropy, demagnetizing, and energy exchange contributions. The magnetization can change in two ways: it can flip suddenly (switching) or turn gradually to follow the magnetic field (rotation). A hard layer around soft grains is created by compressive stress. Future research should focus on investigating the secondary peak in permeability and how the angle alters between hard and soft areas.

The article "***Can We Still Find an Ideal Memristor***?" by F. Z. Wang sheds light on the role of a memristor. An ideal memristor was invented by Chua in 1971 [3]. It correlates

https://doi.org/10.3390/magnetism6010002

the magnetic flux Φ and the electric charge q with the two physical quantities nonlinearly correlated. The direct interaction between Φ and q behaves like a memristor. It is shown that the "ideal/real/perfect" memristor requires magnetism, but the structure still faces two major limitations: (1) an inductive effect, and (2) bistability along with dynamic sweeping across a continuous resistance range. Several innovations based on an "open" structure, nanoscale dimensions, magnetic materials with cubic anisotropy and sequential switching of magnetic domains are proposed. This discovery paves the way for next-generation ideal memristors, even though they can still be considered a purely a mathematical concept.

L. Kong et al. focus on "***Loss Mitigation in Self-Biased Microstrip Circulators***". The use of ferrite devices in radio frequency (RF) systems and antennas is quite challenging because they need external magnets. Hexaferrite-based devices can work without magnets at high frequencies, but the loss of energy is recorded at lower frequencies used in wireless communication. This paper examines La-Co (lanthanum-cobalt)-substituted hexaferrite materials for low-GHz circulators.

A. K. Paul deals with the "***Practical Study of Mixed-Core High Frequency Power Transformer***". In the core of power electronics, the use of medium to high frequency aims to remove heat in order to facilitate optimal use of both the core and copper. The thermal conduction of a transformer device is complicated by the presence of loss centers. By taking the core made by a soft magnetic material, a mixed-core profile to achieve superior temperature profile and better long-term performance is employed. The new cores replace the original ones under the windings. In addition, they share more of the heat dissipation load. The magnetic and thermal behavior of the transformer is assessed through the lens of two high-power applications. In the first case, the core operates at its rated flux, while in the second case, the windings operate at their rated currents.

G. Diguet et al. delve into the realm of "***Magneto Elasticity Modeling for Stress Sensors***". In this article, the phenomena of stress/stress sensing hosted in the thin films is discussed and demonstrated. Different main pillars such as the application of stress (compressive and tensile) and the application of a DC electromagnetic field are the bases of the sensing phenomena. Physical phenomena—magnetic susceptibility and magnetic permeability—are examined by applying a strong exciting AC electromagnetic field aligned along the same direction of the applied stress. The magnetic susceptibility of the material changes with the applied stress and reaches a maximum value when the stress is close to a critical level. For the DC electromagnetic field applied along the direction of the exciting AC electromagnetic field and along the same direction as the applied stress, the critical stress is shifted toward negative values, and for the DC electromagnetic field applied perpendicularly, the critical stress is shifted towards larger positive values. Specifically, when the DC electromagnetic field is in the same direction as the AC one, the critical stress becomes smaller, while the critical stress becomes larger when the two fields are perpendicular. In summary, in this investigation experimental endeavors were conducted to obtain a clear overview on magnetoelastic properties for stress sensors.

B. Kidd proposes "***Vector-Based Magnetic Circuit Modelling of Induction Motors***". An electro-mechanical system is used together with applied and rotating magnetic fields. Efficiency and accuracy are the two main factors needed to create a robust device. The developed vector-based magnetic circuit equations give more accurate results to describe the operational characteristics of induction motors if compared to regular flux models. The exact equations are extracted through the shape and design of the device and the results are reliable with a short calculation time.

M. Ortner et al. propose "***Numerically Stable and Computationally Efficient Expression for the Magnetic Field of a Current Loop***". In this work, the authors show that direct

https://doi.org/10.3390/magnetism6010002

implementations of standard textbook formulas for the off-axis magnetic field of a current loop become numerically unstable over a wide region. Near the axis of symmetry and at large distances from the loop, accuracy deteriorates rapidly. The instability comes from numerical cancellation, which cannot be fixed algebraically. All exact expressions reported in the literature suffer from similar issues. The authors explore unconventional computing compared to conventional methods. Different approaches such as dipole, Taylor, and binomial approaches are compared with classical ones. The novel method is based on an analytic expression using Bulirsch's complete elliptic integral.

N. Maniotis et al. propose ***"The Role of Blood Perfusion in the Thermal Interaction Between Magnetic Nanoparticles and Cancerous Tumors: A Computational Study"***. The model studies how blood flow affects tumor heating during treatment with magnetic nanoparticles. Blood flow removes heat from the tumor, so the temperature increase is reduced. The heat produced by 15 nm magnetite nanoparticles is calculated from micromagnetic simulations. In the numerical simulations are use a magnetic field of about 20 mT and a frequency of 100 kHz. The energy loss of nanoparticles is used to model the tumor and healthy tissue. Normal and cancer tissues are distinguished based on experimental differences in how blood flow changes with temperature. 2D bioheat model is used with nanoparticles evenly spread in the tumor. The results show that tumors retain more heat than healthy tissue because their blood flow increases less under heating, which helps therapy. The tumor reaches a temperature ranging between 41 °C and 45 °C, while the other nearby tissues are unaffected. As a main result, tumor heating treatments are presented and improved through nanoparticles and blood flow.

3. Future Directions

In our opinion, there are numerous promising avenues for future technology based on magnetism. First, in the framework of advanced magnetic phenomena and theoretical insights, magnetic hopfions, which are 3D topological spin-textures, are highly promising, with potential applications in the field of spintronics, memory data storage, and neuromorphic computing. Research focused on magnetic materials, devices, and applications presents future smart materials and potential sensing applications that are built on magnetoelasticity modeling. However, not all of these smart materials and sensing applications rely exclusively on magnetoelastic effects. Some are based on other magnetic phenomena and mechanisms beyond magnetoelasticity. Another main endeavor is based on magnetic modeling and computational methods, mainly related to the efficiency and accuracy of electromechanical systems, and focuses on linking analytical and numerical approaches for real-time applications. The analysis is not limited to electromechanical systems but it can also be applied to other magnetic and multi-physics systems. We pose two open questions which arise as a continuation of the topics discussed in this Special Issue: What about quantum computing linked to magnetism? And what about the synergy between theory and experiment? At the moment, there are no final answers to these questions.

Funding: This research received no funding.

Acknowledgments: We would like to thank all the researchers who submitted their papers to this Special Issue. We congratulate the authors of the published papers and thank them for sharing their excellent results through our platform. Our appreciation also goes to the reviewers who carefully, responsibly, and fairly selected excellent papers for this Special Issue and provided valuable review comments for the authors. We acknowledge the Editorial Board of *Magnetism* for granting us the opportunity to guest edit this Special Issue. Lastly, we are grateful to the *Magnetism* Editorial Office

https://doi.org/10.3390/magnetism6010002

staff for their strict supervision and responsible management, which ensured the timely publication of this Special Issue.

Conflicts of Interest: The authors declare no conflicts of interest.

List of Contributions

1. Hasičić, M.; Angelopoulos, S.; Ktena, A.; Hristoforou, E. A Study on the Effect of Plastic Strain on Magnetic Phenomenology and Microstructure. *Magnetism* **2025**, *5*, 1. https://doi.org/10.3390/magnetism5010001.
2. Maniotis, N.; Mitropoulos, S.; Vordos, N.; Tsiantos, V. The Role of Blood Perfusion in the Thermal Interaction Between Magnetic Nanoparticles and Cancerous Tumors: A Computational Study. *Magnetism* **2025**, *5*, 6. https://doi.org/10.3390/magnetism5010006.
3. Guslienko, K. Magnetic Hopfions: A Review. *Magnetism* **2024**, *4*, 383–399. https://doi.org/10.3390/magnetism4040025.
4. Wang, F.Z. Can We Still Find an Ideal Memristor? *Magnetism* **2024**, *4*, 200–208. https://doi.org/10.3390/magnetism4030014.
5. Kong, L.; Schuchinsky, A.; Joseph, S.; Eker, T.; Huang, Y. Loss Mitigation in Self-Biased Microstrip Circulators. *Magnetism* **2023**, *3*, 121–134. https://doi.org/10.3390/magnetism3020010.
6. Ortner, M.; Leitner, P.; Slanovc, F. Numerically Stable and Computationally Efficient Expression for the Magnetic Field of a Current Loop. *Magnetism* **2023**, *3*, 11–31. https://doi.org/10.3390/magnetism3010002.
7. Paul, A.K. Practical Study of Mixed-Core High Frequency Power Transformer. *Magnetism* **2022**, *2*, 306–327. https://doi.org/10.3390/magnetism2030022.
8. Diguet, G.; Froemel, J.; Kurita, H.; Narita, F.; Makabe, K.; Ohtaka, K. Magneto Elasticity Modeling for Stress Sensors. *Magnetism* **2022**, *2*, 288–305. https://doi.org/10.3390/magnetism2030021.
9. Kidd, B. Vector-Based Magnetic Circuit Modelling of Induction Motors. *Magnetism* **2022**, *2*, 130–151. https://doi.org/10.3390/magnetism2020010.

References

1. Vansteenkiste, A.; Leliaert, J.; Dvornik, M.; Helsen, M.; Garcia-Sanchez, F.; Van Waeyenberge, B. The design and verification of MuMax3. *AIP Adv.* **2014**, *4*, 107133. [CrossRef]
2. Donahue, M.J.; Porter, D.G. *OOMMF User's Guide, Version 1.0*; Interagency Report NISTIR 6376; National Institute of Standards and Technology: Gaithersburg, MD, USA, 1999.
3. Chua, L.O. Memristor—The missing circuit element. *IEEE Trans. Circuit Theory* **1971**, *18*, 507. [CrossRef]

magnetism

MDPI

Article

The Role of Blood Perfusion in the Thermal Interaction Between Magnetic Nanoparticles and Cancerous Tumors: A Computational Study

Nikolaos Maniotis [1,*], Spyridon Mitropoulos [2], Nikolaos Vordos [3] and Vassilios Tsiantos [3]

1 School of Physics, Faculty of Sciences, Aristotle University, 54124 Thessaloniki, Greece
2 Department of Biomedical Engineering, University of West Attica, 12243 Athens, Greece; smitro@uniwa.gr
3 Department of Physics, Democritus University of Thrace, 65404 Kavala, Greece; nvordos@physics.duth.gr (N.V.); tsianto@physics.duth.gr (V.T.)
* Correspondence: nimaniot@physics.auth.gr

Abstract: In this study, the role of blood perfusion in modulating the thermal response of tumors during magnetic nanoparticle hyperthermia was investigated through computational modeling. The thermal dissipation of 15 nm magnetite nanoparticles was estimated using micromagnetic simulations of their hysteresis loops under a magnetic field of 20 mT and a frequency of 100 kHz. These calculations provided precise energy loss parameters, serving as inputs to simulate the temperature distribution in a tumor embedded within healthy tissue. Temperature-dependent blood perfusion rates, derived from experimental models, were integrated to differentiate the vascular dynamics in normal and cancerous tissues. The simulations were conducted using a bioheat transfer model on a 2D axisymmetric tumor geometry with magnetite nanoparticles dispersed uniformly in the tumor volume. Results showed that tumor tissues exhibit limited blood perfusion enhancement under hyperthermic conditions compared to healthy tissues, leading to localized heat retention favorable for therapeutic purposes. The computational framework validated these findings by achieving therapeutic tumor temperatures (41–45 °C) without significant overheating of surrounding healthy tissues, highlighting the critical interplay between perfusion and energy dissipation. These results demonstrate the efficacy of combining nanoparticle modeling with temperature-dependent perfusion for optimizing magnetic nanoparticle-based hyperthermia protocols.

Keywords: magnetic nanoparticles; magnetic nanoparticle hyperthermia; blood perfusion; finite element method; micromagnetic simulation; hysteresis losses; bioheat transfer

1. Introduction

During the clinical application of magnetic nanoparticle hyperthermia (MNH), it is desirable to have complete knowledge of the temperature distribution in order to achieve the therapeutic tumor temperature range (41–45 °C) [1]. Invasive temperature measurements can provide information on the tissue temperature at discrete points only when the number of probes that can be implanted is restricted by patient tolerance and practical aspects, and thus the tissue temperature knowledge is limited [2]. Therefore, a mathematical model for calculating tissue temperature during the treatment could prove valuable in complementing the invasive temperature measurements. The difference between applied (cause) and heat-generated (effect) fields, experimentally seen in vitro and in vivo, is overwhelming, and so is the heat distribution [3]. Bridging theory and experiment in these cases requires new efforts in terms of heat measurement techniques and simulation approaches. Once a match between the latter two has been achieved, there will be real chances of the routine use of MNH as a treatment adapted to the specific requirements of each cancer case. One of the main shortcomings of many in vitro models is the disregard of both tumor and body physiological conditions. For the sake of simplicity, biological media are usually considered to be a blood-flooded matrix composed of cells and interstitial space in physical models of hyperthermia [4,5].

Specific heat, thermal, and electrical conductivities, mass density, as well as the dielectric constant of the involved biological media (tumor and healthy tissue) must also be included in models and simulations. In the case of deep-seated tumors, the perturbing effects of bones (with low dielectric constant and thermal conductivity) are usually neglected due to the added complexity and their relatively limited influence on the heat transfer [6,7]. Another important element is vascularization. The heat transfer process greatly depends on blood perfusion, which is different for tumors and normal tissue. Moreover, bifurcations in vessels have an impact on the cooling effect of blood [8]. Finally, the heat sources, i.e., MNPs, must be included so as to complete the basic tumor model of MNH. These may exhibit size distribution, which influences their magnetic properties and, therefore, the heat generation process, and spatial distribution, which determines the formation of "hot spots" and the uniformity of heat deposition in the tissue where MNPs are infused [9,10].

Heat transfer in living tissues is a complicated process because it involves a combination of thermal conduction in tissues, convection and perfusion of blood, and metabolic heat production [11–13]. Over the years, several mathematical models have been developed to describe heat transfer within living biological tissues. These models have been widely used in the analysis of hyperthermia in cancer treatment, laser surgery, cryosurgery, cryopreservation, thermal comfort, and many other applications. The most widely used bioheat model was introduced by Pennes in 1948 [14]. Pennes proposed a new simplified bioheat model to describe the effect of blood perfusion and metabolic heat generation on heat transfer within a living tissue. Since the publication of Pennes' landmark paper, the bioheat transfer equation (BHTE) of his model has been widely used by many researchers for the analysis of bioheat transfer phenomena. For any MNP construct, thermal modeling is critical for understanding the heating efficacy at cellular level. The BHTE can be used to predict temperature profiles during local hyperthermia [15,16]. Despite the development of more accurate temperature prediction models, the BHTE can still be applied to almost every case of thermal modeling [16–20]. When applying the BHTE, the specified magnetic field strength, frequency, background temperature, estimated average perfusion, MNP concentration, and distribution in the target tissue yield an energy term which is then used to estimate the power absorption and temperature distribution within the target region [21–23].

Blood perfusion can be the dominant form of energy removal when considering heating processes. It assumes that the blood enters the control volume at some arterial temperature and then comes to equilibrium at tissue temperature. Thus, as the blood leaves the control volume, it carries away the energy and therefore acts as an energy sink in MPH treatment. In many works, blood perfusion is considered as a constant value that is characteristic of the tissue [11,24–26]. In reality, blood perfusion is not static; it varies with temperature and time. Tumor microvasculature exhibits unique thermal responses compared to healthy tissues, with an initial hyperperfusion that may decline under prolonged hyperthermic conditions due to vascular collapse or functional impairments [27–29]. Incorporating such complexities is crucial for accurately simulating the thermal interaction between magnetic nanoparticles and tumors. Here, we address this gap by integrating temperature- and time-dependent blood perfusion models into a computational framework for MNH simulations.

This study estimated the thermal dissipation of 15 nm magnetite nanoparticles subjected to an alternating magnetic field of 20 mT and 100 kHz using micromagnetic simulations, providing the energy loss parameters necessary for the analysis. Power dissipation was integrated into a bioheat transfer model, and the finite element method (FEM) was employed to solve the heat transfer equations and compute spatial and temporal temperature distributions within a spherical tumor surrounded by healthy tissue. The blood perfusion rates in both tumor and healthy tissues were modeled as dynamic functions of temperature and time, capturing the physiological variability inherent in the process. Prior to the application of MPH in patients, safety and effectiveness must be ensured with the generation of an accurate and reliable treatment plan. Consequently, the need for a numerical model that considers variables such as the tissue density, specific heat and thermal conductivity, metabolic heat generation, and blood perfusion rate is of high importance. Our model takes into account all the suitable boundary conditions and, more importantly, should be utilized within the framework of BHTE tailored to varied perfusion. Accurate treatment planning can allow for the generation of a temperature spatiotemporal distribution which can provide better visualization of the heat transfer present in the tumor region than single-point thermometry. Thermal modeling that is in agreement with physical temperature measurements can complement focal thermal monitoring. By incorporating temperature-dependent and time-dependent blood perfusion rates into the FEM framework, this work delivers a more accurate representation of thermal dynamics during MNH, offering critical insights into optimizing this therapy for effective and safe clinical applications.

2. Materials and Methods

To determine the hysteresis losses of magnetic nanoparticles (MNPs), a detailed micromagnetic numerical model was developed with OOMMF software [30], assuming single-domain, spheroidal particles with identical material composition. The particles interact with one another via magnetostatic dipole–dipole interactions, which were incorporated into the simulation framework. The model simulates the behaviour of an ensemble of MNPs by numerically integrating the Landau–Lifshitz–Gilbert (LLG) [31] equation to calculate their magnetic response under an alternating magnetic field. The energy losses Q were evaluated by calculating the hysteresis loop area and multiplying it by the applied magnetic field frequency. For these simulations, the Fe_3O_4 MNPs were characterized by the following magnetic parameters: a damping coefficient alpha = 0.02, a saturation magnetization of bulk magnetite Ms = 480 kA/m, and a first-order cubic magnetocrystalline anisotropy constant $K = -13.5\ kJ/m^3$. Those values were chosen based on the study [32] which investigates a system of MNPs with the same properties, including size, shape, material composition, and anisotropy. This ensures that the selected value is representative of the specific MNPs used in our work. The volume concentration of MNPs in the simulated environment was set to $\phi = 1\%$, and the anisotropy axes of individual particles were randomly oriented to mimic realistic spatial distributions. This is an acceptable concentration value usually employed both clinically [33–35] and in silico [36]. Simulations were performed at an initial temperature of 300 K to ensure a thermally consistent representation of the system. Field amplitude and frequency were chosen to simulate realistic clinical conditions [9,37] and were set to 20 mT and 100 kHz, respectively.

The finite element model (FEM) utilized in this study comprises a spherical tumor with a radius of 5 mm embedded within a healthy brain tissue region with a radius of 25 mm. This geometric configuration was adopted from the works of K. Maier-Hauff et al. [38,39].

The dimensions align with the realistic anatomical parameters observed in glioblastoma multiforme cases, enhancing the applicability of the model for clinically relevant scenarios. This configuration enables an accurate representation of heat distribution within the tumor and surrounding healthy tissue during magnetic nanoparticle hyperthermia. In order to save computational time and space but simultaneously obtain accurate results for the realistic 3D space, we designed our geometry in a 2D (r,z) axisymmetric model as depicted in Figure 1. The use of a 2D axisymmetric model is a well-established approach in heat transfer studies, as it significantly reduces computational costs while maintaining the key characteristics of heat distribution [40,41]. This approach is particularly valid when considering radially symmetric tumor geometries and homogeneous heating conditions.

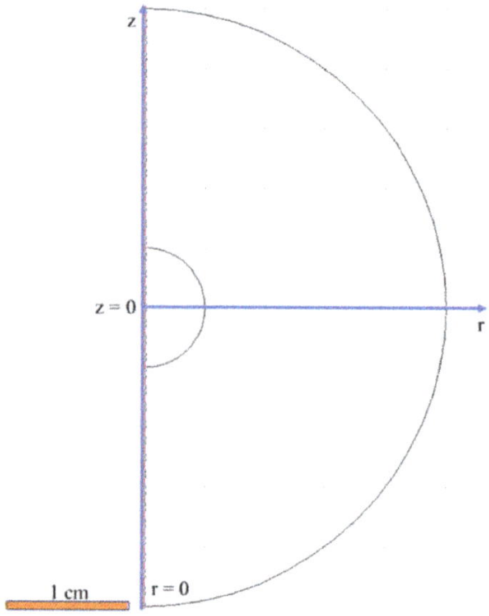

Figure 1. Geometry of the applied model: the model represents both MNPs bearing tumor and healthy tissue as semicircles of radius 5 mm and 25 mm, respectively. The modeling plane is the rz-plane: the horizontal axis represents the r-axis, and the vertical axis represents the z- axis. To obtain the actual 3D geometry, we have to revolve the 2D geometry about the z-axis. The scale bar is also indicated.

The heat transfer simulation was performed using COMSOL 3.5a Multiphysics software [42]. The time-dependent "Bioheat Transfer" module is utilized to simulate the in vivo MNH scheme studied here. The governing equation reads:

$$\rho_t c_t \frac{\partial T}{\partial t} + \nabla \cdot (-k_t \nabla T) = c_b \rho_b \omega_b (T_b - T) + Q_{met} + Q_{ext} \tag{1}$$

where ρ_t, c_t, T, and k_t are density, specific heat, temperature, and thermal conductivity of the tissue, respectively. In addition, T_b is the temperature of arterial blood, Q_{met} is the metabolic heat generation in W/m^3, and ρ_b, c_b, and ω_b are density, specific heat, and perfusion rate of blood, respectively. Finally, Q_{ext} is the energy deposition rate in W/m^3 which is generated by an external heating source (in our case, the MNPs subjected to AC field) and is related to the MNPs' hysteresis losses Q through:

$$Q_{ext} = \phi Q \tag{2}$$

Metabolic heat generation Q_{met} is usually very small compared to the external power deposition term Q_{ext}. It should be noted that metabolic heat generation is assumed to be homogeneously distributed throughout the tissue. The first term on the left part of Equation (1) describes the storage of thermal energy while the first term on the right is related to heat diffusion. The term $\rho_b c_b \omega_b (T_b - T)$, which accounts for the effects of blood perfusion, can be the dominant form of energy removal when considering heating processes. It assumes that the blood enters the control volume at some arterial temperature, T_b, and then reaches equilibrium at tissue temperature. Thus, as the blood leaves the control volume, it carries away the energy and therefore acts as an energy sink in MPH treatment. Power deposition due to eddy currents can be neglected due to the low amplitude and frequency of the applied alternating magnetic field (20 mT/100 kHz).

The thermophysical and other properties of the tumor and healthy tissue used for simulation are summarized in Table 1. The properties of blood are summarized in Table 2.

Table 1. Thermophysical properties of healthy tissue and tumors taken from [43,44].

Tissue	Thermal Conductivity (W/mK) k_t	Density (kg/m^3) ρ_t	Specific Heat (J/kgK) c_t	Viscosity (Pas) n	Metabolic Heat Rate (W/m^3) Q_{met}
Healthy tissue	0.64	1040	3693	0.1	6374.5
Tumor	0.57	1000	3500	0.1	31,872.5

Table 2. Thermophysical properties of blood taken from [32].

	Density (kg/m^3) ρ_b	Specific Heat (J/kgK) c_b	Temperature (°C) T_b
Blood	1060	3894	37

The simulations also adopt the temperature-dependent blood perfusion rate model introduced by Lang et al. [45], for both the tumor, $\omega_{b,tumor}$, and the healthy tissue ($\omega_{b,tissue}$) in (1/s). Those blood perfusion rates depend on the temperature in the following way:

$$\omega_{b,tumor} = \begin{cases} 0.000833 \text{ for } T < 37\ °C \\ 0.000833 - \dfrac{(T-37)^{4.8}}{5438000} \text{ for } 37\ °C \leq T \leq 42\ °C\ | \\ 0.000416 \text{ for } T > 42\ °C \end{cases}$$
$$\omega_{b,tissue} = \begin{cases} 0.00045 + 0.00355 \exp\left(-\dfrac{(T-45)^2}{12}\right) \text{ for } T \leq 45\ °C \\ 0.004 \text{ for } T > 45\ °C \end{cases} \quad (3)$$

Under the "Initial Values" node, the normal body temperature of 37 °C (310.15 K) was set throughout the entire geometry. A general "Heat Source" Q_{ext} was added in the Tumor–MNPs composite domain, assuming that all the MNPs are uniformly dispersed within the tumor region, with a volume fraction of ϕ. A "Temperature" node of constant temperature (Dirichlet boundary condition type) was selected to describe the conditions on the outer boundary of the healthy tissue. Assuming that a sufficient volume of healthy tissue spans between the heat source and the boundary, the metabolism and perfusion mechanisms should counterbalance the effect of the source, thus the temperature at the boundary should be approximately equal to the normal body temperature of 37 °C. Treatment time was equal to 3600 s. At tumor boundaries, we set the continuity boundary condition. This boundary condition describes the energy transfer between the boundaries of the tumor and the external environment, which is the healthy tissue. The general relation that describes this is the following:

$$\mathbf{n} \cdot (\mathrm{k}_1 \nabla \mathrm{T}_1 - \mathrm{k}_2 \nabla \mathrm{T}_2) = 0 \quad (4)$$

where **n** is the vector perpendicular to the boundary surface. This condition specifies that the heat flux in the normal direction is continuous across the boundary between material 1 (tumor) and material 2 (healthy tissue). The axial symmetry boundary condition is applied along the *z*-axis of the model. This condition is used in boundaries in order to reduce the overall size of the model and save system resources like CPU time and space. The boundaries of the model are depicted in Figure 2a with different colored arrows.

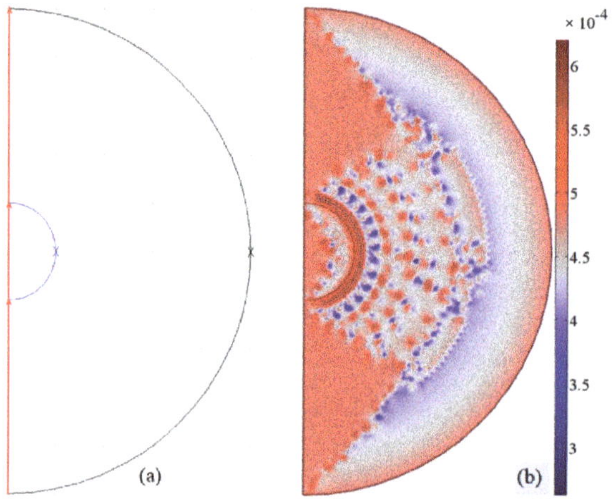

Figure 2. (**a**) Color-indexed arrows for the various boundary conditions of the bioheat transfer problem. (**b**) Size color legend (in meters) of the various triangular mesh elements used for the model.

Since the model contains rotational symmetry, the computational domain can be represented in 2D using cylindrical coordinates. Because of the computation time saved by modeling in 2D, we can select a fine mesh that will provide extremely accurate results. An extra-fine mesh was used throughout the entire geometry as shown in Figure 2b. It consisted of 11,813 triangular domain elements and 586 boundary elements. The time-dependent study of 3600 s is carried out for a time step of 0.1 s and a relative tolerance of 0.001. A direct time-dependent solver is used throughout the simulations that included 23,886 degrees of freedom (DOFs). All the boundary conditions used are presented in Table 3.

Table 3. Simulating conditions used for each boundary.

Arrows Color Index in Figure 2a	Boundary Condition
red	Axial symmetry
blue	Continuity (Energy Transfer)
black	Dirichlet (constant temperature at 37 °C)

3. Results

3.1. Micromagnetic Simulation

The hysteresis loop, depicted in Figure 3, reflects the reduction in magnetization due to the nanoscale size of the particles compared to the bulk material [46]. Furthermore, the low field of 20 mT cannot cause a completely saturated magnetization of the system of interacting nanoparticles since the characteristic field of the cubic anisotropy is as high as 2K/M_s = 60 mT, which is much higher than the coercive field obtained here ($\approx$8 mT). By integrating the data of Figure 3, we found that hysteresis loop area equals to 3.6 J/kg_{MNPs}. By multiplying this value to the driving frequency value, we found $Q_{ext} = 360\ W/g_{MNPs}$.

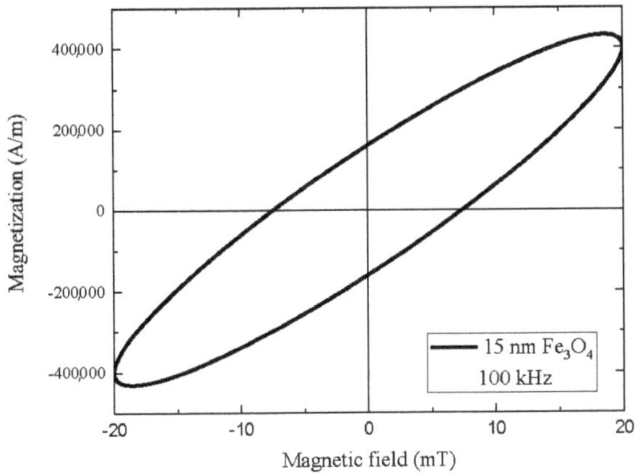

Figure 3. Hysteresis loop for interacting magnetite nanoparticles of diameter 15 nm. The applied magnetic field amplitude and the bulk saturation magnetization are equal to 20 mT and 480 kA/m. Temperature fluctuations also attenuate the magnetization of particles.

3.2. Blood Perfusion

After using COMSOL 3.5a to numerically solve the time-dependent Equation (1), we obtained the distributions of variables such as blood perfusion, temperature, and MNPs' volumetric power dissipation in order to appraise the validity of our method. First, we obtained the blood perfusion temporal distribution for the healthy tissue and tumor which is depicted in Figures 4a and 4b, respectively. Note that the blood perfusion rate distributions presented in Figure 4 correspond to average values across the boundaries of both the tumor and healthy tissue. This approach accounts for spatial variations in temperature and provides a representative measure of the perfusion response in each region.

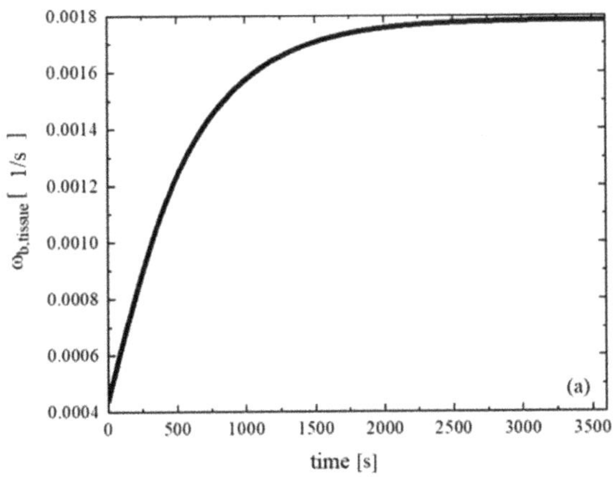

Figure 4. *Cont.*

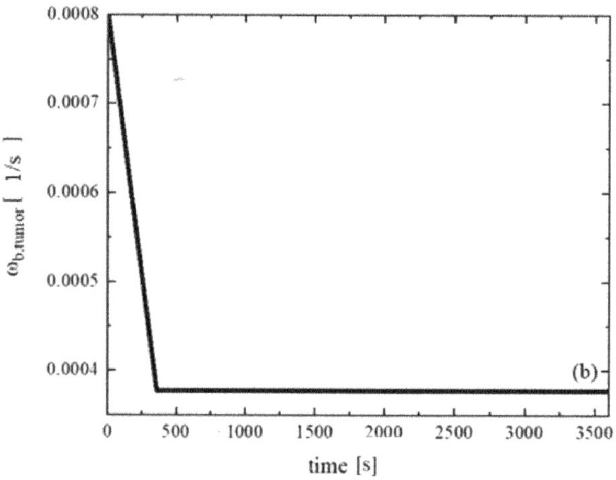

Figure 4. Blood perfusion temporal distribution for (**a**) healthy tissue and (**b**) tumor after 1 h of magnetic nanoparticle hyperthermia (MNH) treatment. The healthy tissue demonstrates an adaptive vascular response, increasing blood flow as temperature rises to enhance heat dissipation. In contrast, the hypertrophic vascular system of the tumor shows an elevated initial blood flow at normal temperatures but a reduced ability to respond at higher temperatures, leading to decreased perfusion. This differential perfusion behavior facilitates selective heating of the tumor, supporting the therapeutic efficacy of MNH.

It can be seen that healthy tissue is characterized by a normal vascular system, which supplies the tissue with more blood as the temperature rises in order to extract heat. The vascular system of the tumor is hypertrophic and consequently, at normal temperature, it provides greater blood flow. On the other hand, at higher temperatures, the vascular system of tumor fails to increase the blood flow and therefore reduces it. Thus, it is easier to heat the tumor. This property is beneficial for the whole process of the in vivo MNH.

3.3. Heat Transfer

Then, we estimated temperature distribution within the geometry. The heating source of the tumor are the MNPs, with their losses given by Equation (2). Since blood acts as a heat sink because of perfusion, an efficient amount of energy is required to attain an MNH-appropriate temperature increase, which is illustrated in Figure S1 after t = 3600 s of MNH treatment. Temperature as a function of the radial distance is depicted in Figure 5, together with the distributions obtained for different values of blood perfusion found in the literature. Figure 5 highlights the significant role of blood perfusion in determining the achievable temperatures within the tumor and surrounding healthy tissue. The curves indicate that higher perfusion rates lead to enhanced heat dissipation, resulting in lower steady-state temperatures. Conversely, lower perfusion rates contribute to increased tumor heating. Our model incorporating variable perfusion $\omega_b = \omega_b(T,t)$ predicts a more realistic temperature distribution, as it accounts for physiological responses to hyperthermia treatment. The results demonstrate that assuming a fixed perfusion rate may underestimate the temperature profile, affecting treatment predictions. This variability emphasizes the need for considering a range of perfusion values, as biological properties are subject to individual differences.

A contour temperature plot within the tumor volume is presented in Figure S2a. We see that the entire volume of the tumor enters the MNH window limits, i.e., between

41 °C and 45 °C, whereas tumor boundaries are above the MNH threshold. On the other hand, the temperature within the healthy tissue did not increase significantly, as illustrated in the contour temperature plot of Figure S2b. The maximum temperature reached in healthy tissue volume is 41 °C, while at the center of the tumor, the temperature reached the value of 45.6 °C. This temperature increase in healthy tissue ($\Delta T = 4$ °C) does not affect cellular mechanisms, as already been demonstrated in the literature [47].

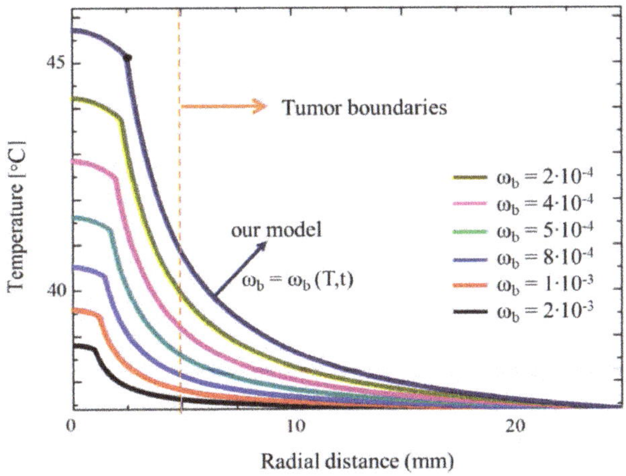

Figure 5. Temperature distribution as a function of the radial distance from the tumor center, incorporating different tumoral blood perfusion rates, ω_b, found in the literature. A vertical dashed line marks the tumor boundary, distinguishing the intratumoral and healthy tissue regions. The plot includes various constant perfusion rates as well as our proposed temperature- and time-dependent perfusion model, demonstrating its impact on thermal behavior. The black point on our model's curve indicates the position where the distribution curvature changes due to high temperature gradients within the tumor.

To validate our method, temperature values attained both in the center and at the boundaries of the tumor were compared to the corresponding experimental temperature values obtained in a clinical studies [38,39] under the same particles (material, size, concentration) and field (amplitude and frequency) conditions. There, the authors performed heating treatment to patients suffering from glioblastoma multiforme by intracranial injection of the magnetic fluid MFL AS (MagForce Nanotechnologies AG, Berlin, Germany [48]), which is subject to European medical device regulations [35] and consisted of aminosilane-coated magnetite nanoparticles (15 nm core size) dispersed in water at the same concentration (1%) used here. Temperature values attained both in the center and at the boundaries of the tumor are compared to the corresponding experimental temperature values obtained in [29,30] to evaluate our method. The results are presented in Table 4 where a good agreement between experimentally measured and numerically calculated values is revealed.

Table 4. Comparison between experimental (Tmeasured) and numerical (Tcalculated) temperature values estimated intratumorally and at tumor boundaries.

Position in Tumor	Tmeasured [29,30] (°C)	Tcalculated (°C)	Deviation (%)
center	44.6	45.6	2.2
boundaries	42	41.5	1.2

Time evolution of temperature at the position where the curvature of their radial distribution changes (2.5 mm from the center of the tumor) and at tumor boundaries (5 mm from the center of the tumor) are depicted in Figure 6.

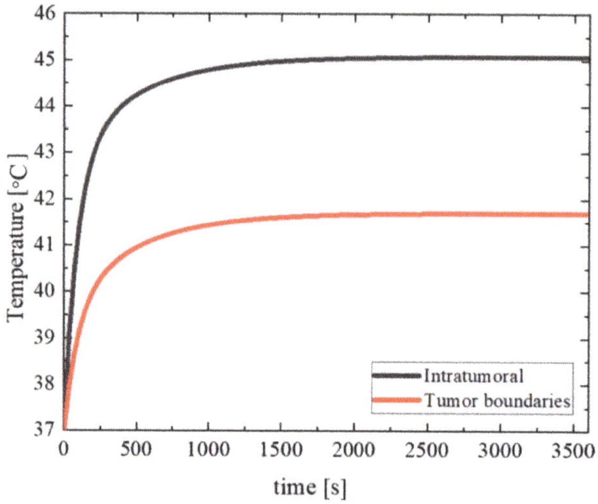

Figure 6. Temperature time evolution at two different tumor positions. The intratumoral distribution corresponds to the site positioned 2.5 mm from the center of the tumor (black point in Figure 5) while the distribution in the boundaries between tumor and healthy tissue is taken at 5 mm.

The results illustrated in Figure 6 show a significant and rapid rise in temperature within the intratumoral region, which stabilizes after approximately 15 min, reaching a steady state. This elevated and sustained temperature is crucial for inducing thermal damage to tumor cells. Conversely, the tumor boundaries exhibit a slower rate of temperature increase and reach a lower steady-state temperature compared to the intratumoral region. This gradient in temperature distribution can be attributed to the heat dissipation effects of surrounding healthy tissues, which have a more robust vascular response and blood perfusion rate. The distinct temperature profile highlights the efficiency of MNH in selectively heating the tumor core while sparing adjacent healthy tissue. Such a differentiation in temperature dynamics between the tumor center and its periphery underscores the importance of the tumor's impaired vascular system, which limits heat dissipation.

Finally, we estimated the Cumulative Equivalent Minutes at 43 °C (CEM43) to quantify the thermal dose delivered during the treatment based on the expression [49,50]

$$\text{CEM43} = \int_{t=0}^{t_{\text{final}}} R^{(43-T(t))} dt \tag{5}$$

where R is a temperature-dependent factor: R = 0.5R = 0.5 for $T \geq 43$ °C and R = 0.25 for T < 43 °C. From temperature profiles in Figure 6, we found that the intratumoral region reaches CEM43 = 60 min within 16 min while the tumor boundary requires 29 min to reach the same dose. These characteristics ensure a therapeutically effective thermal dose to the tumor while minimizing potential thermal damage to the surrounding tissue. These results confirm that the applied heating strategy provides an effective thermal dose in accordance with established hyperthermia treatment standards.

4. Discussion

This study investigates the thermal interaction between magnetic nanoparticles (MNPs) and a cancerous tumor using computational models to evaluate the impact of temperature- and time-dependent blood perfusion rates. The findings illustrate the complex interplay between nanoparticle heat dissipation, tissue perfusion, and tumor physiology, providing critical insights into the optimization of magnetic nanoparticle hyperthermia (MNH) for clinical cancer treatment. The results confirm that tumor tissues can be selectively heated during MNH due to their impaired vascular response, which contrasts with the robust blood perfusion system in healthy tissues. The tumor's hypertrophic vascular structure initially exhibits elevated blood perfusion compared to normal tissue. However, its inability to maintain this increase under hyperthermic conditions results in reduced heat dissipation, allowing for sustained temperature rise in the tumor core. These findings are consistent with previous studies [3,11,51,52] that highlight the vulnerability of tumor vasculature to thermal stress, which enhances the therapeutic efficacy of hyperthermia treatments.

Micromagnetic simulations of 15 nm magnetite nanoparticles were used to accurately determine their thermal dissipation, incorporating realistic material properties such as saturation magnetization, damping coefficient, and anisotropy. In the temperature range of room temperature ($\Delta T \approx 20$ K), the heating efficiency of MNPs remains largely unaffected, as the Curie temperature of magnetite ($\approx$850 K) is significantly higher, ensuring stable magnetic properties and negligible variations in the specific loss power (SLP) during hyperthermia treatment. The power dissipation derived from these simulations matches well with earlier studies using similar nanoparticle systems [53–57]. The adoption of temperature- and time-dependent blood perfusion models in finite element simulations further advances the state-of-the-art by capturing dynamic physiological responses, addressing a critical limitation in previous literature that relied on static perfusion rates. In conventional models, blood perfusion is treated as a fixed parameter characteristic of specific tissue types, disregarding its physiological response to temperature changes. However, in reality, perfusion varies with temperature, particularly in tumor tissues where vascular regulation is impaired. The assumption of a constant perfusion rate leads to an underestimation of the tumor's temperature rise during hyperthermia. This could result in an inaccurate assessment of treatment efficacy, as the thermal dose delivered to the tumor would appear lower than what actually occurred. In contrast, our model accounts for the reduction in tumor perfusion at higher temperatures, better reflecting the real physiological response and predicting a more localized and sustained heating effect. The implications of this discrepancy are critical for hyperthermia treatment planning. Underestimating the temperature increase could lead to suboptimal treatment protocols, potentially reducing therapeutic effectiveness. Conversely, an overestimation of heat dissipation in healthy tissues could cause unnecessary concerns regarding thermal side effects. Our findings highlight that incorporating a variable perfusion model is essential for improving the accuracy of MNH simulations, ensuring a more reliable prediction of the thermal response and enhancing the precision of clinical hyperthermia applications. Furthermore, the inclusion of parametric analysis strengthens the study by providing a broader perspective on treatment effectiveness under different physiological conditions.

From the bioheat transfer simulations, it is evident that the MNH treatment achieves therapeutic temperatures (41–45 °C) in the tumor without overheating the surrounding healthy tissue. This selective heating highlights the potential of MNH in minimizing adverse side effects, which is a persistent challenge in other cancer treatment modalities like chemotherapy and external beam radiotherapy. The use of realistic anatomical dimensions for the tumor and brain geometry, based on the work of Maier-Hauff et al. [29,30], ensures clinical relevance and applicability of the results to glioblastoma multiforme treatment scenarios. The findings also endorse the working hypothesis that the differential thermal and vascular responses of tumors and healthy tissues can be exploited to achieve precise and localized heating of tumors during MNH. In our model, we assume that the heat generated by MNPs via hysteresis losses is uniformly distributed within the tumor volume. This assumption is based on the use of surfactants and other stabilizing agents used in the literature that help ensure a homogeneous dispersion of MNPs, thereby minimizing aggregation effects. Variations in MNP concentration may occur due to aggregation driven by strong dipolar interactions. These spatial inhomogeneities could lead to localized differences in heating efficiency. Our recent study [58] has demonstrated that MNP concentration significantly affects heating efficiency, as evidenced by specific loss power (SLP) estimations derived from micromagnetic simulations. In scenarios where aggregation occurs, the overall heating efficiency might be reduced, potentially leading to deviations in the predicted temperature profiles. Thus, several factors require further investigation. For example, nanoparticle distribution within the tumor and agglomeration effects could influence the overall heat generation, warranting more advanced modeling techniques. Similarly, the impacts of non-uniform tissue properties and variations in vascular density need to be explored to refine the predictive accuracy of computational models.

Future research should focus on validating these computational findings with experimental in vivo studies to establish robust correlations between modeling predictions and biological outcomes. Investigating the long-term effects of MNH treatment, including tumor recurrence and vascular recovery, would provide a more comprehensive understanding of its therapeutic potential. Furthermore, the combination of MNH with other treatment modalities, such as immunotherapy or targeted drug delivery, could open new avenues for enhancing therapeutic outcomes.

5. Conclusions

By employing the experimental conditions and MNP parameters that were used in a clinical application of MNH found in the literature, bioheat transfer modeling was carried out in order to visualize the temperature profile within the biological tissues (tumor and healthy tissue). Simulations revealed that temperature increase and heating rate within the tumor are maintained in ideal therapeutic levels (within moderate hyperthermia levels) without affecting thermally the healthy tissue. The results are also verified in the literature and validate our technique.

The modeling of heat generation and transfer process in MNH will increase the understanding of physical phenomena and allow a successful transition of this technology from bench to bedside. In addition, modeling through simulation can be used in the planning of the MPH treatment and it also serves as a new alternative method for temperature mapping due to the difficulty in real-time clinical temperature measurement during the treatment. Conclusively, the beneficial role of modeling implementation in an in vivo MPH scheme can be illustrated through the following aspects:

(i) MPH treatment can be optimized with respect to the heat application geometry by maximizing the therapeutic effect while minimizing unwanted side effects.
(ii) The outcome of the treatment can be evaluated based on model predictions.
(iii) It can also be used for extensive parametric studies in order to characterize the stability of various treatment parameters and conditions.
(iv) New treatment strategies can be proposed and evaluated.

Supplementary Materials: The following supporting information can be downloaded at: https://www.mdpi.com/article/10.3390/magnetism5010006/s1, Figure S1: Temperature mapping within our model geometry after 3600 s of MPH treatment at 20 mT/100 kHz due to the applied volumetric power dissipation in tumor; Figure S2: Temperature contour plots within (**a**) tumor and (**b**) healthy tissue.

Author Contributions: Conceptualization, N.M.; methodology, N.M.; software, N.M. and S.M.; validation, N.M., V.T. and S.M.; formal analysis, N.M.; investigation, N.M., V.T., N.V. and S.M.; resources, N.M., V.T. and S.M.; data curation, N.M., V.T., S.M. and N.V.; writing—original draft preparation, N.M.; writing—review and editing, N.M., V.T., S.M. and N.V.; visualization, N.M., V.T. and S.M.; supervision, N.M. and V.T.; project administration, N.M. and V.T. All authors have read and agreed to the published version of the manuscript.

Funding: This research received no external funding.

Institutional Review Board Statement: Not applicable.

Informed Consent Statement: Not applicable.

Data Availability Statement: The data presented in this study are available in this article.

Conflicts of Interest: The authors declare no conflicts of interest.

References

1. Johannsen, M.; Gneveckow, U.; Taymoorian, K.; Thiesen, B.; Waldöfner, N.; Scholz, R.; Jung, K.; Jordan, A.; Wust, P.; Loening, S.A. Morbidity and Quality of Life during Thermotherapy Using Magnetic Nanoparticles in Locally Recurrent Prostate Cancer: Results of a Prospective Phase I Trial. *Int. J. Hyperth.* **2007**, *23*, 315–323. [CrossRef] [PubMed]
2. Saville, S.L.; Qi, B.; Baker, J.; Stone, R.; Camley, R.E.; Livesey, K.L.; Ye, L.; Crawford, T.M.; Thompson Mefford, O. The Formation of Linear Aggregates in Magnetic Hyperthermia: Implications on Specific Absorption Rate and Magnetic Anisotropy. *J. Colloid Interface Sci.* **2014**, *424*, 141–151. [CrossRef] [PubMed]
3. Hedayatnasab, Z.; Abnisa, F.; Daud, W.M.A.W. Review on Magnetic Nanoparticles for Magnetic Nanofluid Hyperthermia Application. *Mater. Des.* **2017**, *123*, 174–196. [CrossRef]
4. Périgo, E.A.; Hemery, G.; Sandre, O.; Ortega, D.; Garaio, E.; Plazaola, F.; Teran, F.J. Fundamentals and Advances in Magnetic Hyperthermia. *Appl. Phys. Rev.* **2015**, *2*, 041302. [CrossRef]
5. Chen, L.; Chen, C.; Wang, P.; Song, T. Mechanisms of Cellular Effects Directly Induced by Magnetic Nanoparticles under Magnetic Fields. *J. Nanomater.* **2017**, *2017*, 1564634. [CrossRef]
6. Romeo, D.; Salieri, B.; Hischier, R.; Nowack, B.; Wick, P. An Integrated Pathway Based on in Vitro Data for the Human Hazard Assessment of Nanomaterials. *Environ. Int.* **2020**, *137*, 105505. [CrossRef]
7. He, S.; Zhang, H.; Liu, Y.; Sun, F.; Yu, X.; Li, X.; Zhang, L.; Wang, L.; Mao, K.; Wang, G.; et al. Maximizing Specific Loss Power for Magnetic Hyperthermia by Hard-Soft Mixed Ferrites. *Small* **2018**, *14*, 1800135. [CrossRef]
8. Yue, K.; Yu, C.; Lei, Q.; Luo, Y.; Zhang, X. Numerical Simulation of Effect of Vessel Bifurcation on Heat Transfer in the Magnetic Fluid Hyperthermia. *Appl. Therm. Eng.* **2014**, *69*, 11–18. [CrossRef]
9. Dutz, S.; Hergt, R. Magnetic Nanoparticle Heating and Heat Transfer on a Microscale: Basic Principles, Realities and Physical Limitations of Hyperthermia for Tumour Therapy. *Int. J. Hyperth.* **2013**, *29*, 790–800. [CrossRef]
10. Dutz, S.; Hergt, R. Magnetic Particle Hyperthermia—A Promising Tumour Therapy? *Nanotechnology* **2014**, *25*, 452001. [CrossRef]
11. Spirou, S.V.; Basini, M.; Lascialfari, A.; Sangregorio, C.; Innocenti, C. Magnetic Hyperthermia and Radiation Therapy: Radiobiological Principles and Current Practice. *Nanomaterials* **2018**, *8*, 401. [CrossRef] [PubMed]
12. Socoliuc, V.; Peddis, D.; Petrenko, V.I.; Avdeev, M.V.; Susan-Resiga, D.; Szabó, T.; Turcu, R.; Tombácz, E.; Vékás, L. Magnetic Nanoparticle Systems for Nanomedicine—A Materials Science Perspective. *Magnetochemistry* **2020**, *6*, 2. [CrossRef]
13. Obaidat, I.M.; Narayanaswamy, V.; Alaabed, S.; Sambasivam, S.; Muralee Gopi, C.V.V. Principles of Magnetic Hyperthermia: A Focus on Using Multifunctional Hybrid Magnetic Nanoparticles. *Magnetochemistry* **2019**, *5*, 67. [CrossRef]
14. Pennes, H.H. Analysis of Tissue and Arterial Blood Temperatures in the Resting Human Forearm. *J. Appl. Physiol.* **1948**, *1*, 93–122. [CrossRef]
15. Giordano, M.A.; Gutierrez, G.; Rinaldi, C. Fundamental Solutions to the Bioheat Equation and Their Application to Magnetic Fluid Hyperthermia. *Int. J. Hyperth.* **2010**, *26*, 475–484. [CrossRef]
16. Bellizzi, G.; Bucci, O.M. On the Optimal Choice of the Exposure Conditions and the Nanoparticle Features in Magnetic Nanoparticle Hyperthermia. *Int. J. Hyperth.* **2010**, *26*, 389–403. [CrossRef]

17. Kandala, S.K.; Liapi, E.; Whitcomb, L.L.; Attaluri, A.; Ivkov, R. Temperature-Controlled Power Modulation Compensates for Heterogeneous Nanoparticle Distributions: A Computational Optimization Analysis for Magnetic Hyperthermia. *Int. J. Hyperth.* **2019**, *36*, 115–129. [CrossRef]
18. Vicentini, M.; Ferrero, R.; Manzin, A. Influence of Coil Geometry, Supply Conditions and Nanoparticle Heating Properties on Magnetic Hyperthermia in Mouse Models. *Int. J. Therm. Sci.* **2024**, *203*, 109151. [CrossRef]
19. Manzin, A.; Ferrero, R.; Vicentini, M. From Micromagnetic to In Silico Modeling of Magnetic Nanodisks for Hyperthermia Applications. *Adv. Theory Simul.* **2021**, *4*, 2100013. [CrossRef]
20. Nabil, M.; Decuzzi, P.; Zunino, P. Modelling Mass and Heat Transfer in Nano-Based Cancer Hyperthermia. *R. Soc. Open Sci.* **2015**, *2*, 150447. [CrossRef]
21. Louis, D.N.; Perry, A.; Reifenberger, G.; von Deimling, A.; Figarella-Branger, D.; Cavenee, W.K.; Ohgaki, H.; Wiestler, O.D.; Kleihues, P.; Ellison, D.W. The 2016 World Health Organization Classification of Tumors of the Central Nervous System: A Summary. *Acta Neuropathol.* **2016**, *131*, 803–820. [CrossRef] [PubMed]
22. Le, T.A.; Bui, M.P.; Yoon, J. Theoretical Analysis for Wireless Magnetothermal Deep Brain Stimulation Using Commercial Nanoparticles. *Int. J. Mol. Sci.* **2019**, *20*, 2873. [CrossRef] [PubMed]
23. Kumar, A.; Attaluri, A.; Mallipudi, R.; Cornejo, C.; Bordelon, D.; Armour, M.; Morua, K.; Deweese, T.L.; Ivkov, R. Method to Reduce Non-Specific Tissue Heating of Small Animals in Solenoid Coils. *Int. J. Hyperth.* **2013**, *29*, 106–120. [CrossRef] [PubMed]
24. Sebeke, L.; Deenen, D.A.; Maljaars, E.; Heijman, E.; de Jager, B.; Heemels, W.P.M.H.; Grüll, H. Model Predictive Control for MR-HIFU-Mediated, Uniform Hyperthermia. *Int. J. Hyperth.* **2019**, *36*, 1039–1049. [CrossRef]
25. Huang, H.W.; Horng, T.L. *Bioheat Transfer and Thermal Heating for Tumor Treatment*; Elsevier Inc.: Amsterdam, The Netherlands, 2015; ISBN 9780124079007.
26. Wu, L.; Cheng, J.; Liu, W.; Chen, X. Numerical Analysis of Electromagnetically Induced Heating and Bioheat Transfer for Magnetic Fluid Hyperthermia. *IEEE Trans. Magn.* **2015**, *51*, 4600204. [CrossRef]
27. Rodrigues, H.F.; Capistrano, G.; Mello, F.M.; Zufelato, N.; Silveira-Lacerda, E.; Bakuzis, A.F. Precise Determination of the Heat Delivery during in Vivo Magnetic Nanoparticle Hyperthermia with Infrared Thermography. *Phys. Med. Biol.* **2017**, *62*, 4062–4082. [CrossRef]
28. Tay, Z.W.; Chandrasekharan, P.; Chiu-Lam, A.; Hensley, D.W.; Dhavalikar, R.; Zhou, X.Y.; Yu, E.Y.; Goodwill, P.W.; Zheng, B.; Rinaldi, C.; et al. Magnetic Particle Imaging-Guided Heating in Vivo Using Gradient Fields for Arbitrary Localization of Magnetic Hyperthermia Therapy. *ACS Nano* **2018**, *12*, 3699–3713. [CrossRef]
29. Suriyanto; Ng, E.Y.K.; Kumar, S.D. Physical Mechanism and Modeling of Heat Generation and Transfer in Magnetic Fluid Hyperthermia through Néelian and Brownian Relaxation: A Review. *Biomed. Eng. Online* **2017**, *16*, 36. [CrossRef]
30. Donahue, M.J.; Porter, D.G. *OOMMF User's Guide*, version 1.0.; National Institute of Standards and Technology: Gaithersburg, MD, USA, 1999. [CrossRef]
31. Sundara Mahalingam, S.; Manikandan, B.V.; Arockiaraj, S. Review—Micromagnetic Simulation Using OOMMF and Experimental Investigations on Nano Composite Magnets. *J. Phys. Conf. Ser.* **2019**, *1172*, 012070. [CrossRef]
32. Ferrero, R.; Vicentini, M.; Manzin, A. Influence of Size, Volume Concentration and Aggregation State on Magnetic Nanoparticle Hyperthermia Properties versus Excitation Conditions. *Nanoscale Adv.* **2024**, *6*, 1739–1749. [CrossRef]
33. Chen, G.; Roy, I.; Yang, C.; Prasad, P.N. Nanochemistry and Nanomedicine for Nanoparticle-Based Diagnostics and Therapy. *Chem. Rev.* **2016**, *116*, 2826–2885. [CrossRef] [PubMed]
34. Bruschi, M.L.; de Toledo, L.d.A.S. Pharmaceutical Applications of Iron-Oxide Magnetic Nanoparticles. *Magnetochemistry* **2019**, *5*, 50. [CrossRef]
35. Rubia-Rodríguez, I.; Santana-Otero, A.; Spassov, S.; Tombácz, E.; Johansson, C.; De La Presa, P.; Teran, F.J.; Morales, M.D.P.; Veintemillas-Verdaguer, S.; Thanh, N.T.K.; et al. Whither Magnetic Hyperthermia? A Tentative Roadmap. *Materials* **2021**, *14*, 706. [CrossRef] [PubMed]
36. Maniotis, N. *Numerical Simulation and Characterization of Systems for Magneto-Thermal and Magneto-Mechanical Interaction with the Living Matter*; Aristotle University of Thessaloniki: Thessaloniki, Greece, 2020.
37. Zeisberger, M.; Dutz, S.; Müller, R.; Hergt, R.; Matoussevitch, N.; Bönnemann, H. Metallic Cobalt Nanoparticles for Heating Applications. *J. Magn. Magn. Mater.* **2007**, *311*, 224–227. [CrossRef]
38. Maier-Hauff, K.; Rothe, R.; Scholz, R.; Gneveckow, U.; Wust, P.; Thiesen, B.; Feussner, A.; Deimling, A.; Waldoefner, N.; Felix, R.; et al. Intracranial Thermotherapy Using Magnetic Nanoparticles Combined with External Beam Radiotherapy: Results of a Feasibility Study on Patients with Glioblastoma Multiforme. *J. Neurooncol.* **2007**, *81*, 53–60. [CrossRef]
39. Maier-Hauff, K.; Ulrich, F.; Nestler, D.; Niehoff, H.; Wust, P.; Thiesen, B.; Orawa, H.; Budach, V.; Jordan, A. Efficacy and Safety of Intratumoral Thermotherapy Using Magnetic Iron-Oxide Nanoparticles Combined with External Beam Radiotherapy on Patients with Recurrent Glioblastoma Multiforme. *J. Neurooncol.* **2011**, *103*, 317–324. [CrossRef]

40. He, J.C.; Bao, Y.; Chen, X. Asymmetry of Two-Dimensional Thermal Convection at High Rayleigh Numbers. *Symmetry* **2024**, *16*, 1583. [CrossRef]
41. Matle, S.; Sundar, S. Axi Symmetric 2D Simulation and Numerical Heat Transfer Characteristics for the Calibration Furnace in a Rectangular Enclosure. *Appl. Math. Model.* **2012**, *36*, 878–893. [CrossRef]
42. *COMSOL Multiphysics®*, version 3.5a; COMSOL, Inc.: Burlington, MA, USA, 1998.
43. Miaskowski, A.; Sawicki, B. Magnetic Fluid Hyperthermia Modeling Based on Phantom Measurements and Realistic Breast Model. *IEEE Trans. Biomed. Eng.* **2013**, *60*, 1806–1813. [CrossRef]
44. Mcintosh, R.L.; Anderson, V. Erratum: "A Comprehensive Tissue Properties Database Provided for the Thermal Assessment of a Human at Rest". *Biophys. Rev. Lett.* **2013**, *08*, 99–100. [CrossRef]
45. Lang, J.; Erdmann, B.; Seebass, M. Impact of Nonlinear Heat Transfer on Temperature Control in Regional Hyperthermia. *IEEE Trans. Biomed. Eng.* **1999**, *46*, 1129–1138. [CrossRef] [PubMed]
46. Kim, T.; Shima, M. Reduced Magnetization in Magnetic Oxide Nanoparticles. *J. Appl. Phys.* **2007**, *101*, 09M516. [CrossRef]
47. Chalkidou, A.; Simeonidis, K.; Angelakeris, M.; Samaras, T.; Martinez-Boubeta, C.; Balcells, L.; Papazisis, K.; Dendrinou-Samara, C.; Kalogirou, O. In Vitro Application of Fe/MgO Nanoparticles as Magnetically Mediated Hyperthermia Agents for Cancer Treatment. *J. Magn. Magn. Mater.* **2011**, *323*, 775–780. [CrossRef]
48. Jordan, A.; Scholz, R.; Maier-Hauff, K.; Johannsen, M.; Wust, P.; Nadobny, J.; Schirra, H.; Schmidt, H.; Deger, S.; Loening, S.; et al. Presentation of a New Magnetic Field Therapy System for the Treatment of Human Solid Tumors with Magnetic Fluid Hyperthermia. *J. Magn. Magn. Mater.* **2001**, *225*, 118–126. [CrossRef]
49. Sapareto, S.A.; Dewey, W.C. Thermal Dose Determination in Cancer Therapy. *Int. J. Radiat. Oncol. Biol. Phys.* **1984**, *10*, 787–800. [CrossRef]
50. van Rhoon, G.C.; Franckena, M.; ten Hagen, T.L.M. A Moderate Thermal Dose Is Sufficient for Effective Free and TSL Based Thermochemotherapy. *Adv. Drug Deliv. Rev.* **2020**, *163–164*, 145–156. [CrossRef]
51. Zhang, E.; Kircher, M.F.; Koch, M.; Eliasson, L.; Goldberg, S.N.; Renström, E. Dynamic Magnetic Fields Remote-Control Apoptosis via Nanoparticle Rotation. *ACS Nano* **2014**, *8*, 3192–3201. [CrossRef]
52. Mody, V.V.; Cox, A.; Shah, S.; Singh, A.; Bevins, W.; Parihar, H. Magnetic Nanoparticle Drug Delivery Systems for Targeting Tumor. *Appl. Nanosci.* **2014**, *4*, 385–392. [CrossRef]
53. Skumiel, A.; Kaczmarek, K.; Flak, D.; Rajnak, M.; Antal, I.; Brząkała, H. The Influence of Magnetic Nanoparticle Concentration with Dextran Polymers in Agar Gel on Heating Efficiency in Magnetic Hyperthermia. *J. Mol. Liq.* **2020**, *304*, 112734. [CrossRef]
54. Lachowicz, D.; Górka, W.; Kmita, A.; Bernasik, A.; Zukrowski, J.; Szczerba, W.; Sikora, M.; Kapusta, C.; Zapotoczny, S. Enhanced Hyperthermic Properties of Biocompatible Zinc Ferrite Nanoparticles with a Charged Polysaccharide Coating. *J. Mater. Chem. B* **2019**, *7*, 2962–2973. [CrossRef]
55. Brzuszek, K.; Ross, C.A.; Janutka, A. High-Frequency Magnetic Response of Superparamagnetic Composites of Spherical Fe and Fe3O4 Nanoparticles. *J. Magn. Magn. Mater.* **2024**, *599*, 172070. [CrossRef]
56. Ashraf, M.; Khan, I.; Usman, M.; Khan, A.; Shah, S.S.; Khan, A.Z.; Saeed, K.; Yaseen, M.; Ehsan, M.F.; Tahir, M.N.; et al. Hematite and Magnetite Nanostructures for Green and Sustainable Energy Harnessing and Environmental Pollution Control: A Review. *Chem. Res. Toxicol.* **2020**, *33*, 1292–1311. [CrossRef] [PubMed]
57. Wu, K.; Wang, J.P. Magnetic Hyperthermia Performance of Magnetite Nanoparticle Assemblies under Different Driving Fields. *AIP Adv.* **2017**, *7*, 056327. [CrossRef]
58. Maniotis, N.; Gitsou, M.; Maragakis, M. Conjugating Magnetic Nanoparticles Anisotropy to Their Dipolar Interactions: Effect on the Hyperthermic Losses Index via Micromagnetic Simulations. *J. Magn. Magn. Mater.* **2025**, *617*, 172843. [CrossRef]

magnetism

MDPI

Article

A Study on the Effect of Plastic Strain on Magnetic Phenomenology and Microstructure

Mehrija Hasičić [1,*], Spyridon Angelopoulos [2], Aphrodite Ktena [2,3] and Evangelos Hristoforou [2]

[1] Department of Electrical and Electronics Engineering, Faculty of Engineering, Natural and Medical Sciences, International Burch University, 71210 Ilidža, Bosnia and Herzegovina
[2] Sensors Laboratory, School of Electrical and Computer Engineering, National Technical University of Athens, 157 72 Zografou, Greece; apktena@uoa.gr (A.K.); hristoforou@ece.ntua.gr (E.H.)
[3] Energy Systems Laboratory, Core Department, National & Kapodistrian University of Athens, 344 00 Euboea, Greece
* Correspondence: mehrija.hasicic@ibu.edu.ba

Abstract: The present work aspires to contribute to the discussion on the relationship between macroscopic measurements and microstructure, helping establish a methodology that will allow the quantitative assessment of the effect of strain on magnetic properties in the plastic deformation regime. In particular, we study the effect of strain on the magnetization process as a result of varying the anisotropy profile at the grain level. Results on micromagnetic calculations of hysteresis loops for various configurations of magnetic anisotropy are shown and discussed against the interplay between the energy terms involved in the calculations, namely anisotropy, demagnetizing, and exchange. The results are in line with previously obtained results using vector Preisach modeling with the Stoner–Wohlfarth model acting both as a switching and rotation mechanism. The hysteresis loop phenomenology is consistent with the emergence of a hard phase in the form of a boundary around soft grains which is assumed to be the result of the onset of compressive stresses in the plastic region. Future research will be oriented toward the study of the effect of the secondary peak in differential permeability, which is observed experimentally in the plastic deformation region, and its dependence on the angle of misalignment between the hard boundary and the soft grain.

Keywords: macroscopic measurements; micromagnetic calculations; hysteresis loops

1. Introduction

Nondestructive testing and evaluation (NDT&E) techniques are important both for the monitoring of critical structures as well as for zero-defect manufacturing. In particular, magnetic NDT&E is one of the most promising technologies where magnetic materials are involved, as in the case of steel structures and the steel manufacturing industry.

The underlying concept of magnetic NDT&E is the dependence of the magnetization process on strain. Mechanical stresses applied at the manufacturing stage, mechanical and thermal treatments, loading, and aging during the lifetime of a material are all causes of strain which manifests itself as deformation at the lattice, grain, or macroscopic level and results in residual stresses [1].

These techniques fall under two major categories, namely those based on major or minor loop measurements [2–6] and those based on magnetic Barkhausen noise [7–11] measurements. Coercivity, remanence, and magnetic permeability are all properties that have been proposed as metrics in the first case, while the noise root mean square (rms) voltage envelope, counts or energy are proposed for the second category.

Magnetic NDT&E techniques are appealing due to the low-cost sensor arrangements and their adaptability to various applications in the field, which is most important for both monitoring and manufacturing uses. However, it often suffers from drawbacks, such as low repeatability as well as a lack of benchmarking and standardization, which makes it appropriate for qualitative rather than quantitative assessment [6,11–14].

Modeling the effect of strain on the magnetization process is important for understanding the underlying mechanisms governing the dependence of the measured/monitored magnetic parameters on microstructural changes related to strain and residual stresses.

The modeling of stress-dependent magnetization processes has been studied by a number of groups at various scales [15–18]. The stress tensor contributes to the effective magnetic field distribution in a strained magnetic material and affects the magnetization reversal mechanism; e.g., when a tensile (positive) stress is applied along the dominant magnetization axis of a sample with positive magnetostriction, it assists the alignment of the domains with the dominant magnetization direction, whereas if compressive (negative) stress is applied, the domains tend to rotate away from the dominant magnetization direction. This approach has been used in a modification of the Jiles–Atherton model [19]. The effect of stress and/or microstructure in order to make better predictions on the efficiency and performance of different electrical machines has been modeled for several applications: the effect of grain size and dislocation density on hysteretic magnetic properties in steels [19], the influence of biaxial stress on anhysteretic behavior [20], strain broadening under the effect of dislocations [21], asymmetry in the magnetomechanical effect under tensile and compressive stress [22], and the effect of multiaxial stress on magnetic hysteresis of electrical steel sheets [23].

The stress-depending effective field approach is not as straightforward in the case of plastically deformed materials in the unloaded state where compressive stresses coexist with tensile stresses and the stress tensor is not known.

In this work, we aspire to contribute to the discussion on the relationship between macroscopic measurements and microstructure on the way toward establishing a methodology that will allow the quantitative assessment of the effect of strain on magnetic properties in the plastic deformation regime. Our starting point is experimental evidence and previous modeling results concerning martensitic steels [15]. Tensile stress in the plastic deformation regime results in the emergence of compressive residual stresses and a magnetically hard phase in the grain boundaries [1,24]. This is reflected on hysteresis loops with increased coercivity as well as lower remanence and differential permeability. This phenomenology has been reproduced by a modeling approach based on the Preisach formalism which assumes the existence of a hard phase and stress-dependent long-range interactions, i.e., magnetostatic energy increasing with strain [15]. In this work, we treat the effect of strain on the magnetization process in the plastic deformation regime as a result of a varying anisotropy profile at the grain level, and more specifically, we consider the emergence of a magnetic hard phase in the form of a grain boundary. Using the open-source software OOMMF 1.2b2 [25,26] and the energy minimization approach, major loops are calculated for various configurations, and the interplay between the energy terms involved is studied.

In the following section, we present the materials and experimental evidence on which our work is based, and we describe the proposed methodology. In Section 3, we present the simulation results and their discussion, and the conclusions are presented in Sections 4 and 5, respectively.

2. Materials and Methods

The assumptions of the modeling approach presented in this work originate from observations on experimental results. In particular, the measurements shown here have been obtained using commercial electrical steel laminates, featuring a low-carbon steel typically used as the magnetic core in transformers and motors and typical of similar measurements in the literature [1,2,4,10,11,14,27]. Samples 30 mm × 3 mm cut from the same 0.5 mm sheet [2] have been used.

Three types of measurements are used to study the response of the sample to applied magnetic fields as well as to applied and residual stresses. First, stress–strain curves (Figure 1) have been obtained at three different rates of applied stress, namely 0.1 mm/min, 0.5 mm/min, and 1 mm/min, using an INSTRON 8800 machine. The stress–strain curve is used to establish the ranges of the elastic and plastic regions of the samples used in this study. The transition to the plastic region is observed at strain levels exceeding 3%.

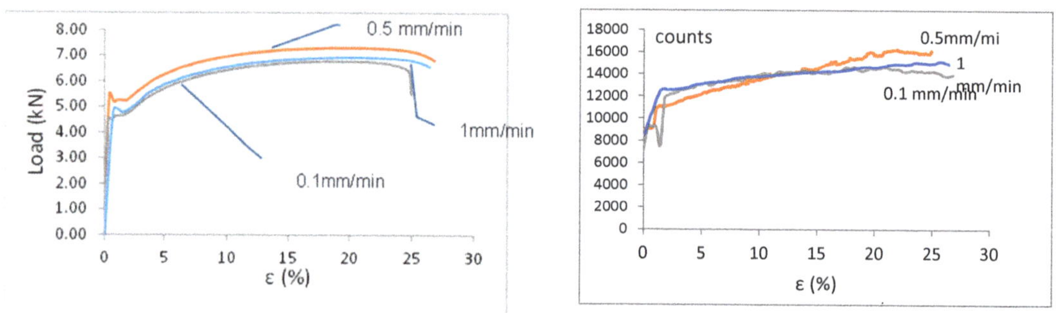

(**a**) (**b**)

Figure 1. (**a**) Stress–strain curves measured at different rates of applied tensile stress and (**b**) corresponding magnetic Barkhausen noise counts.

During the stress–strain curve measurement, a sensor monitoring the magnetic Barkhausen noise (mBN), as described in [26], was also attached to the sample. The excitation is provided by a triangular waveform at 10 Hz. The magnetic field is applied along the same direction as the applied stress. The mBN counts (using MagLab MEB2c) exceeding a specific threshold are used as the mBN metric of choice [11] in Figure 1. Similar results are obtained when the mBN voltage is used instead. The effect applied stress on mBN counts is directly linked to the effect of stress on strain. This measurement demonstrates the coupling between magnetic and mechanical properties and the potential of magnetic macroscopic parameters for the monitoring of mechanical properties such as strain. The transition to the plastic region at strains higher than 3% is also observed in the mBN measurements.

The third type of measurement concerns hysteresis loop measurements using AC magnetometry where an excitation coil wrapped around a yoke [26] generates a low frequency < 0.5 Hz sinusoidal field and magnetizes the sample [2,6,11]. A sensing coil wound around the sample measures the induced output voltage which is proportional to the differential permeability of the material in the volume enclosed by the sensing coil. The hysteresis loops (Figure 2) are obtained on unloaded strained samples to study the effect of residual stress on the magnetization process; first, tensile stress is applied on the sample using the 0.5 mm/min strain rate up to a given strain level. Then, the sample is unloaded, and the hysteresis loop is measured. The magnetic field is applied along the direction the stress has been previously applied, namely along the length of the sample. The hysteresiograph is an inhouse device which has not been calibrated, hence the use a.u.

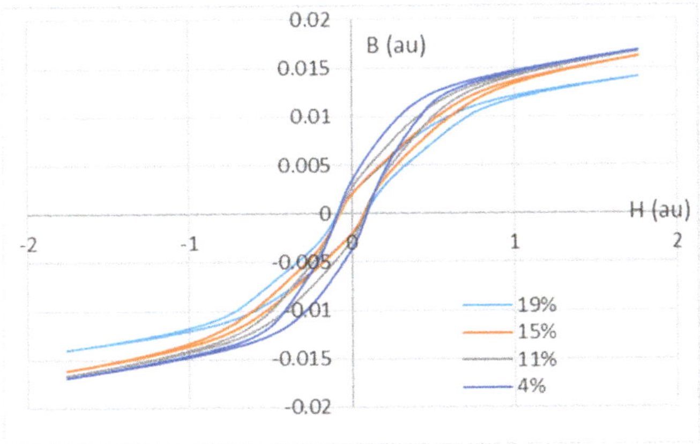

Figure 2. Hysteresis loops measured at various strain levels.

In the plastic deformation regime, there is a monotonic dependence of coercivity and differential permeability on strain. Coercivity increases and permeability decreases. At the microstructural level, higher plastic strain is related to more, finer and misaligned grains and the emergence and development of a hard phase around soft grains, which is the result of the compressive residual stresses that set in [1,24]. The coercivity increase is explained as a result of the hard phase which increases with strain in the plastic region and increased domain wall pinning due to misaligned grains and longer grain boundaries. The decrease in differential permeability is attributed to the emergence of hard boundaries, as the result of compressive residual stresses, whose effective anisotropy is transverse or not aligned with the direction of tensile stress which caused the plastic deformation.

In our previous work, a Preisach model was constructed which used the Stoner–Wohlfarth mechanism as a 2D vector hysteresis operator allowing for both reversible and irreversible rotation and assumed a dispersion of anisotropy around a main easy direction. The emergence of the hard magnetic phase was introduced via secondary Preisach density functions and long-range interactions [15].

Based on the conclusions drawn from both experimental and modeling results, we use the OOMMF open-source software [25] to model the effect of microstructure on the hysteresis loop computed as the result of the free energy minimization for each applied field value. The total energy for a given applied field value is the sum of the Zeeman, anisotropy, exchange, and magnetostatic energy terms.

The reason for this approach is that in engineering applications, the measured parameters involved in the NDT&E of systems containing magnetic materials are usually those obtained from a major loop: e.g., coercivity, remanence, differential permeability, etc. As explained in the Introduction, the goal is to eventually correlate macroscopic parameters obtained from major loop measurements to microstructural changes related to strain. The effect of strain is modeled indirectly through microstructural configurations involving more than one magnetic phase.

In order to study the effect of microstructure and magnetic parameters on the phenomenology depicted by the major loop calculation, the following approach has been designed.

The assumptions of the simulated experiments are outlined below:

- The sample studied (Figure 3) is a parallelpiped of dimensions $L \times L \times H\ (\mathrm{nm})^3$. L is in the order of 2000 nm and H is at least two orders of magnitude smaller.

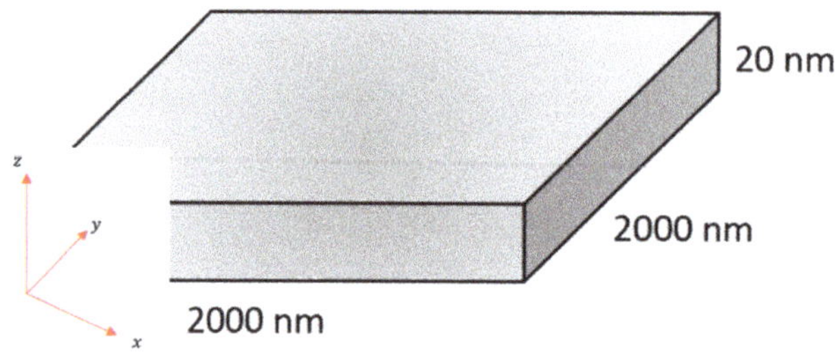

Figure 3. Geometry of a sample modeled.

- It is discretized in N cubic cells of side a nm. N is determined by the sample dimensions and the cell size by dividing each side of the sample with the parameter a. The cell size depends on the magnetostatic exchange length for the soft magnetic materials and monocrystalline exchange length for the hard magnetic materials, which are determined by the material parameters M_S, K_1, and/or A_{ex}, namely, the saturation magnetization, anisotropy constant and exchange constant, respectively.
- Each cell represents one magnetic dipole with parameters M_S, K_1, and/or A_{ex}. Figure 4 shows typical magnetization plots generated by the software of a soft magnetic matrix and soft grains enclosed by misaligned hard boundaries.

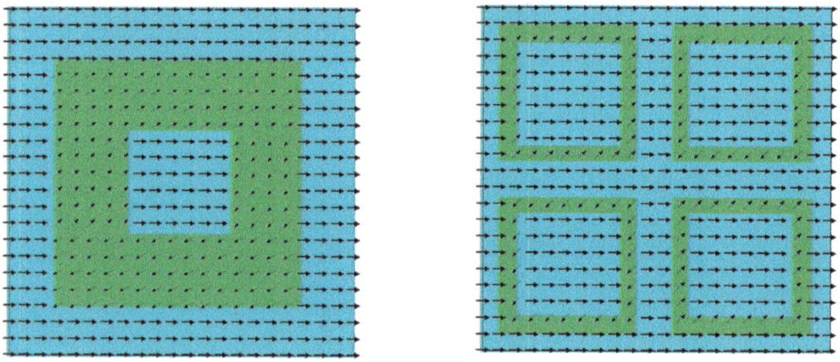

(**a**) (**b**)

Figure 4. Magnetization plots for two types of simulated samples: (**a**) soft grain with hard magnetic boundary and (**b**) four identical soft grains surrounded by misaligned hard magnetic boundaries. The total volume of hard magnetic material is the same in both samples. For clarity, one arrow represents the magnetization of 19 neighboring cells.

Several simulation experiments have been performed to study the effect of microstructure on the major loop, e.g., hard inclusions in a soft matrix, hard inclusions and voids, and finally soft grains surrounded by misaligned hard boundaries [13–15] as those depicted in Figure 4, which is closer to what is reported in experimental studies. The geometries of the models presented in Figure 4a,b serve as initial conditions of the simulations presented in Table 1, (#2–#5) and (#6–#11), respectively. Material parameters were changed to study the effects on the hysteresis loop as explained later in the paper and presented in Table 1.

Table 1. Simulation parameters of the configurations shown in Figure 4.

#	Configuration	K_1—Magnitude [J/m^3]	K_1—Direction [x y z]	A_{ex}, [J/m]	M_S—Magnitude [A/m]	Applied Field [mT] [x y z]
1	Homogeneous case	Ignore	1 0 0	1.30×10^{-11}	1.70×10^{6}	[1000 0 0]
2	4 hard grain boundaries	4.80×10^{3} 1.00×10^{6}	1 0 0 1 1 0	1.30×10^{-11} 1.30×10^{-11}	1.70×10^{6} 1.40×10^{6}	[1000 1000 0]
3	4 hard grain boundaries	4.80×10^{3} 1.00×10^{6}	1 0 0 1 1 0	1.30×10^{-11} 1.30×10^{-11}	1.70×10^{6} 1.40×10^{6}	[1000 0 0]
4	4 hard grain boundaries	4.80×10^{3} 1.00×10^{6}	1 0 0 1 1 0	2.10×10^{-11} 1.30×10^{-11}	1.70×10^{6} 1.40×10^{6}	[1000 1000 0]
5	4 hard grain boundaries	4.80×10^{3} 1.00×10^{6}	1 0 0 1 1 0	2.10×10^{-11} 2.10×10^{-11}	1.70×10^{6} 1.40×10^{6}	[1000 1000 0]
6	1 hard (thin) grain boundary	4.80×10^{3} 1.00×10^{6}	1 0 0 1 1 0	1.30×10^{-11} 1.30×10^{-11}	1.70×10^{6} 1.40×10^{6}	[1000 1000 0]
7	1 hard (thick) grain boundary: 179200 hard cells	4.80×10^{3} 1.00×10^{6}	1 0 0 1 1 0	1.30×10^{-11} 1.30×10^{-11}	1.70×10^{6} 1.40×10^{6}	[1000 1000 0]
8	1 hard (thick) grain boundary: 179200 hard cells	4.80×10^{3} 5.00×10^{5}	1 0 0 1 1 0	1.30×10^{-11} 1.30×10^{-11}	1.70×10^{6} 1.40×10^{6}	[1000 1000 0]
9	1 hard grain boundary _179200hardcells_topbottomYdir	4.80×10^{3} 1.00×10^{6}	1 0 0 1 1 0 0 1 0	1.30×10^{-11} 1.30×10^{-11}	1.70×10^{6} 1.40×10^{6}	[1000 1000 0]
10	1 hard grain boundary _179200hard-cells_topbottomstrongerx	4.80×10^{3} 1.00×10^{6}	1 0 0 1 1 0 1 0.5 0	1.30×10^{-11} 1.30×10^{-11}	1.70×10^{6} 1.40×10^{6}	[1000 1000 0]
11	1 hard grain boundary _179200hard-cells_topbottomstrongerx_Hxdir	4.80×10^{3} 1.00×10^{6}	1 0 0 1 1 0 1 0.5 0	1.30×10^{-11} 1.30×10^{-11}	1.70×10^{6} 1.40×10^{6}	[1000 0 0]

In the energy minimization approach, the magnetization process is the result of the interplay between the energy terms involved, namely the exchange, anisotropy and magnetostatic energy densities, and the applied field. Therefore, we focus on the effect of microstructural parameters on the energy terms involved.

3. Results

The results reported here concern the geometry shown in Figure 4. The simulation parameters for these results are summarized in Table 1. In the column "Configuration", we present different simulations studies: homogeneous case, different cases that include four soft grains (#2–#5) with hard boundaries in a soft matrix ("4 hard grain boundaries"), and the cases with one soft grain (#6–#11) with hard boundaries in a soft matrix ("1 hard (thick) grain boundary"). The presented case with the four soft grains is then examined more through changing the value of the exchange energy coefficient (A [pJ/m]) and the direction of the applied field to observe the effect of the exchange energy coefficient and the applied field on the hysteresis. Then, to observe the effect of the geometry of the soft grains, we have simulated a sample with only one soft grain with hard boundaries where we observed one scenario when the hard boundary is thin (#6), and the rest of the cases are with thick boundaries (#6–#11) keeping the same number of the hard cells in the grain boundary (179,200 hard cells) as used in (#2–#5). Furthermore, we used different anisotropy profiles in the top and bottom part of the boundary by changing the direction of the anisotropy in those sections (#9–#11), while the sides of the hard grain boundary remain the same. The direction of the anisotropy in the mentioned cases is indicated in the name of the configuration; i.e., (#9) "1 hard grain boundary _179200hardcells_topbottomYdir" indicates that the simulated sample has one soft grain with hard boundaries that contains 179,200 hard cells, and the top and bottom of the grain have anisotropy along the positive y-direction. The anisotropy profile of the whole sample can be read from columns K_1-magnitude and K_1-direction, where the first value in the cell represents the magnitude/direction of the anisotropy in the soft matrix and the grain, while the second value represents the magnitude/direction of the anisotropy in the grain boundary. In the case where we have two entries in the K_1-direction cell for the hard grain boundary, the first one represents the anisotropy direction in the sides of the grain boundary, while the second one represents the anisotropy direction of the top and bottom of the grain boundary, i.e., (#9). Here, we have indicated that the top and bottom are in the positive y-direction, as discussed above, the K_1-direction cell has values 1 0 0, 1 1 0, and 0 1 0, which means that the soft matrix and the grain have anisotropy along the positive x-direction [1 0 0], the sides (left and right) of the grain boundary have anisotropy along the xy-plane with 45 degrees being the angle of the direction [1 1 0], and the top and bottom are oriented along the positive y-direction [0 1 0].

4. Discussion

Figure 5 shows the comparison of major hysteresis loops obtained for cases #6, #7 and #8 with one soft grain (Figure 4a) surrounded by a hard magnetic boundary of variable thickness and anisotropy direction against the hysteresis loop of soft homogeneous magnetic material which represents the base case (#1), which we consider the 'unstrained' case. The increase in boundary thickness corresponds to higher levels of plastic deformation which is related to higher compressive residual stresses.

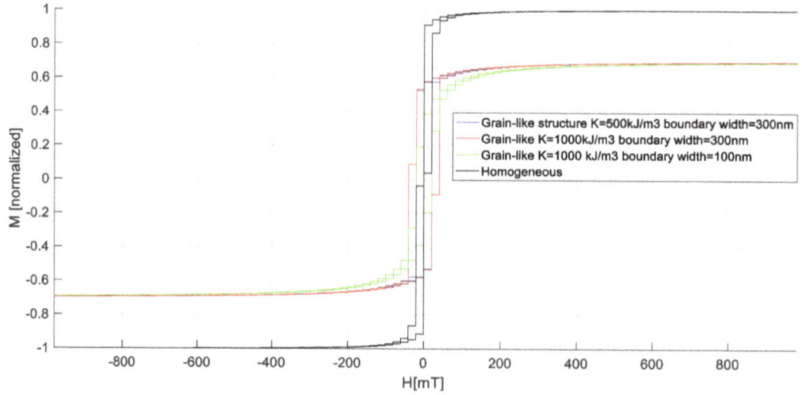

Figure 5. Major hysteresis loop obtained for the soft homogeneous case #1 (black), case #6 with one soft grain enclosed by a hard grain boundary of $K_1 = 1000$ kJ/m^3 and 100 nm thickness (green), case #7 with one soft grain enclosed by a hard grain boundary with $K_1 = 1000$ kJ/m^3 and 300 nm thickness (red) and case #8 with soft grain enclosed by a hard grain boundary with $K_1 = 500$ kJ/m^3 and 300 nm thickness (blue).

The misaligned hard grain boundary leads to a major hysteresis loop with lower saturation since a number of grains can no longer align with the field. This is in line with experimental evidence (Figure 2). The maximum value of saturation reached depends on the direction of the anisotropy of the hard boundary. In the results shown in Figure 5, the direction of the anisotropy and applied field direction is kept the same. A thinner grain boundary yields lower remanence for the same anisotropy constant. However, the observed coercivity depends both on the thickness of the hard boundary or the volume of the hard phase and the anisotropy constant. The sample with the thicker and harder boundary yielded the highest coercivity though not the lowest remanence. Experimental evidence reports on the non-monotonic dependence of remanence on strain, which is the reason why it is not a preferred quantity to monitor in magnetic NDT&E techniques.

Next, we show the effect of the studied microstructure on the energy terms involved in the calculations. The interplay of the four energy components for the four loops shown in Figure 5 is presented in Figure 6a–d.

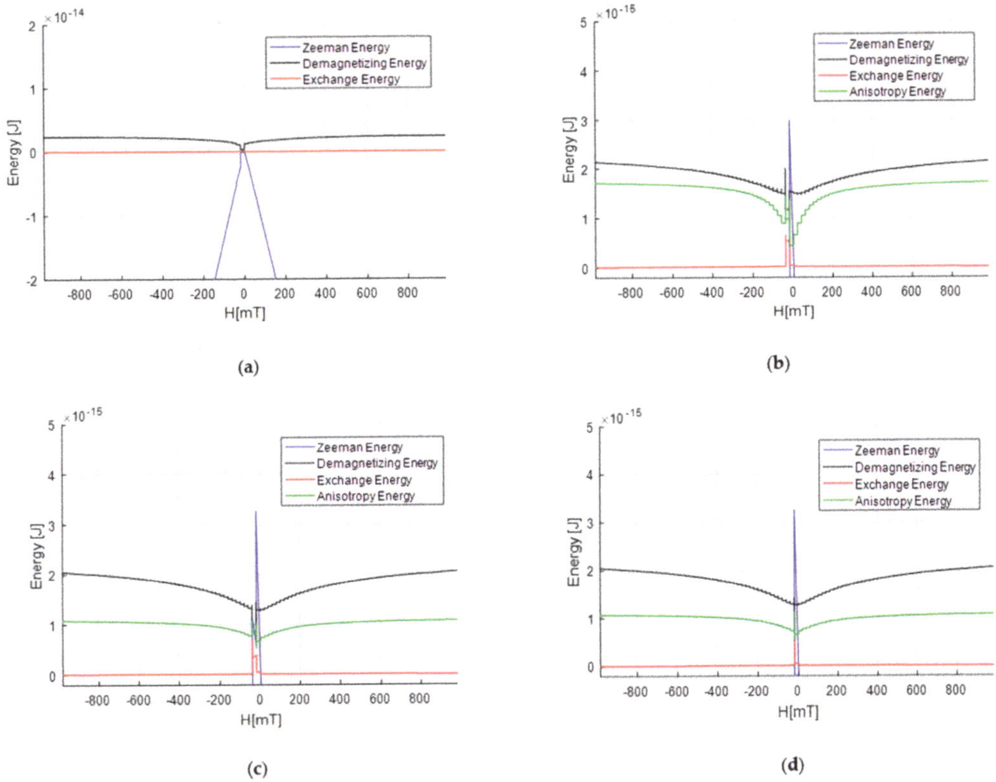

Figure 6. Energy terms of the (**a**) homogeneous case #1, (**b**) case #6 with hard grain boundary of K_1 = 1000 kJ/m^3 and 100 nm thickness, (**c**) case #7 with hard grain boundary of K_1 = 1000 kJ/m^3 and 300 nm thickness and (**d**) case #8 with hard grain boundary of K_1 = 500 kJ/m^3 and 300 nm thickness.

The shape and parameters of the major hysteresis loop are mainly determined by the interplay between the anisotropy and magnetostatic energy. In all cases shown in Figure 6b–d, the magnetostatic energy has the same value approaching saturation, which is in line with the saturation magnetization being the same for the same orientation of the hard phase earlier discussed. On the other hand, the value of the anisotropy energy term toward saturation is higher when the boundary thickness is lower, but its slope increases as the field approaches zero. This is in line with the lower remanence observed in Figure 5 when the boundary is thinner. Both anisotropy and magnetostatic energy are gradually decreasing as the applied field decreases from saturation to allow for reversible rotations away from the initial orientation direction. However, when the sample enters the magnetization reversal region, we observe a steeper decrease in the anisotropy energy with a sharp increase before coercivity which resists the reversal. The profile of this peak in anisotropy energy as well as in the exchange energy controls the coercivity value and depends on both the thickness and the hardness of the boundary. Exchange energy does not seem to have major effect, and we observe a peak in the exchange energy only at coercivity. The Zeeman energy term increases slightly with the boundary thickness.

Next, we explore the effect of anisotropy orientation in the hard grain boundary. We study the case of one hard grain boundary, case #10, where the anisotropy of the top and bottom sides is pinned in the y-direction ([0 1 0]), while on the right and left sides, it remains along the xy-drection, ±45 degrees ([1 1 0]), as shown in case #8. The results obtained are compared against the loop obtained for the configuration shown in Figure 4a, case #8, and presented in Figures 7 and 8.

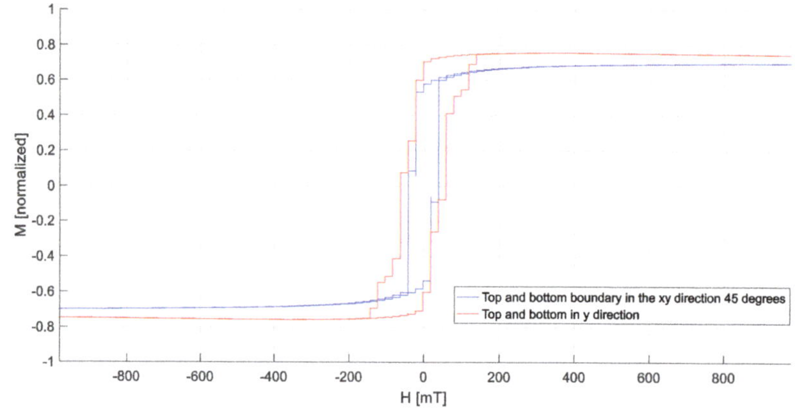

Figure 7. Major hysteresis loop for the case #7 of one hard grain boundary with K_1 = 1000 kJ/m^3 in xy-direction ±45 degrees and 300 nm thickness (blue) compared against the case #9 of a hard grain boundary with K_1 = 1000 kJ/m^3 and 300 nm thickness where the top and bottom side are pinned along the y-direction while the left and right side are kept in the xy-drection ±45 degrees (red).

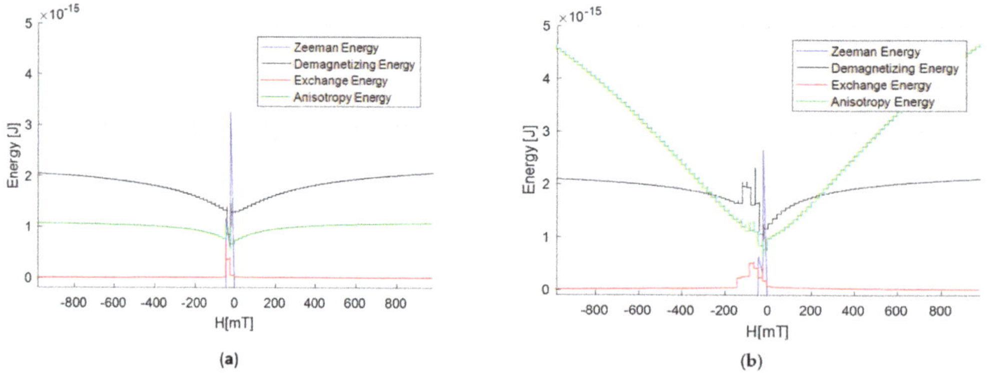

Figure 8. Energy terms for the loops shown in Figure 7: (**a**) hard grain boundary of 300 nm thickness with K_1 = 1000 kJ/m^3 in the xy-direction ±45 degrees and (**b**) hard grain boundary of 300 nm thickness with K_1 = 1000 kJ/m^3 where the top and bottom sides are oriented along the y-direction while the left and right sides are kept along the xy-drection ±45 degrees).

Pinning one part of the hard grain boundary along the *y*-axis affected all of the energy terms involved. It resulted in significantly stronger anisotropy energy which yielded a higher coercivity as observed in the major loop shown in Figure 7. However, as the field decreases from saturation, the anisotropy decreases almost linearly, which is consistent with rotation mainly of the soft phase and the high remanence observed. Both magnetostatic and exchange energy terms control the magnetization reversal, while the Zeeman term is less enhanced. The switching of the soft matrix around the grain occurs first. Vortexes are formed at the edges of the configuration studied, moving from the top and bottom of the *xy plane* and meeting in the middle of the material. The soft matrix having smaller anisotropy switches first, while the hard boundaries due to the stronger anisotropy cannot align with the field, which results in decreased remanence and increased coercivity. The soft material inside the grain switches at a lower field—however, not at the same one as the rest of the soft material.

Next, in Figure 9 we compare the loops obtained for the configurations of cases #7 and #2 shown in Figure 4a,b. The volume of hard phase is the same in both cases; however, in case #7 of Figure 4a, there is one soft grain surrounded by a hard boundary, while in case #2 of Figure 4b, there are four soft grains enclosed by hard boundaries. All other parameters and anisotropy orientations are kept the same in both cases.

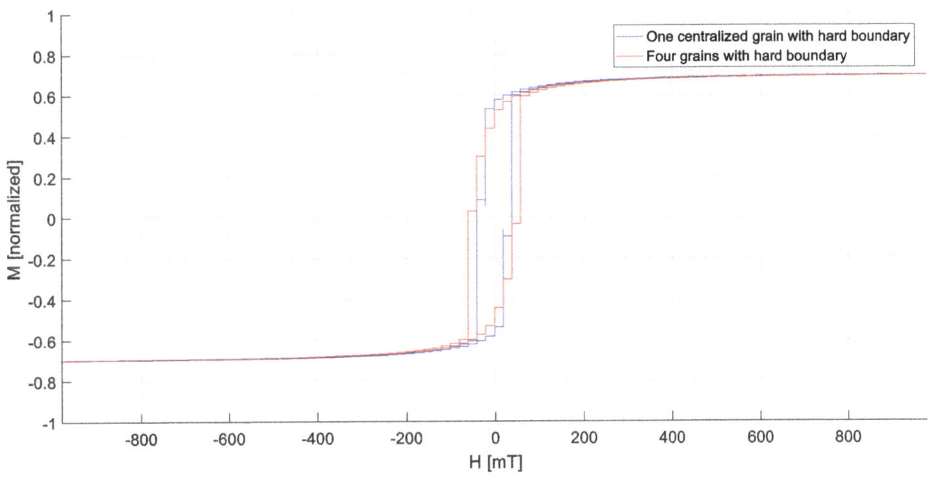

Figure 9. Major hysteresis loop obtained for case #7 with one soft grain with hard grain boundary (blue) and case #2 with four soft grains with hard grain boundary (red).

The two loops in Figure 9 have the same saturation since the anisotropy orientation of the hard boundaries is the same in both cases. For the four-grain configuration, the remanence is slightly lower and the slope of the loop, e.g., the differential permeability, is lower, which is in line with the increasing and higher slope of the anisotropy energy term as the field approaches zero (Figure 10b). Coercivity is higher and controlled by the profile of the magnetostatic and exchange energy. The peaks of all four energy terms are higher at coercivity.

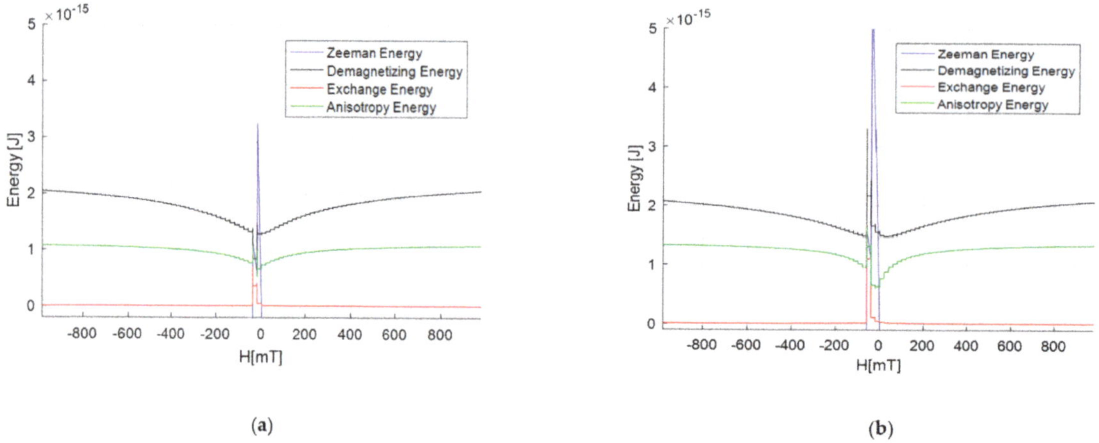

Figure 10. Energy terms for (**a**) case #7 with one soft grain with hard grain boundary and (**b**) case #2 with four soft grains with hard grain boundary.

Finally, looking closer into the values of the various energy terms, we found that the maximum values of demagnetizing and exchange energy were higher in the studies performed on the four-grain configuration. The maximum value of the anisotropy on average is higher in the studies performed on the four-grain model as well; however, the highest value of the anisotropy energy among all studies has been observed in the on-grain configuration when the top and bottom of the hard grain boundary had an enhanced x component.

Overall, the results corroborate our initial assumption that in the plastic deformation region, the magnetization configuration is the result of the emergence of a magnetically hard phase which is the result of the onset of compressive residual stresses which increase with strain. Also, the effect of plastic strain on the macroscopic magnetic parameters depends on the angle of misalignment between the hard boundary and the soft grain, which affects the demagnetizing and anisotropy energy.

5. Conclusions

In this work, we use micromagnetic calculations with an open-source software to study the effect of microstructure on the phenomenology of the major hysteresis loop. The motivation stems from the need to link the macroscopic parameters obtained from magnetic measurements used in NDT&E, such as the differential permeability and coercivity, to plastic strain levels. The main assumption of our calculations is that strain in the plastic region leads to compressive residual stresses which are linked to the emergence of a hard magnetic phase in the form of a grain boundary. The major loop calculations were based on energy minimization involving four energy terms: namely anisotropy, exchange, magnetostatic and Zeeman terms. In our calculations, we focused on the properties of this boundary and studied the resulting interplay of the four energy terms.

The simulations have been carried out for two different configurations: one with one grain enclosed by hard boundaries and the other with four symmetrical grains enclosed by hard boundaries while keeping the same the amount of the hard material in the model. The effect of the anisotropy profile in each configuration was explored, and the results showed that for the same amount of the hard material in the model, lower remanence and higher coercivity were observed when the model was composed of more grains. All energy terms peak around the coercivity field while the anisotropy energy is increasing as the switching occurs, which is contrary to what we observed in the interplay of the energy terms obtained on the configuration with four soft grains enclosed by hard magnetic boundaries.

The effect of texture was investigated through the change in the anisotropy profile of the hard magnetic boundary. The top and bottom of the hard magnetic boundary were pinned in the y-direction while the rest of the configuration was kept the same. Lower remanence, higher coercivity and lower permeability were observed in this case. This is consistent with the assumption about the onset of compressive stresses in the plastic region.

The effect of the amount of the hard material and the effect of the magnitude of the anisotropy of the hard phase was explored using the configuration with one soft grain enclosed by a hard magnetic boundary with various thickness and anisotropy constants. The results were in line with what was observed experimentally; i.e., less hard material or a decrease in the anisotropy of the hard phase both result in lower coercivity. This is consistent with the assumption that plastic stresses favor the emergence of the hard phase.

Future research will be oriented toward the study of the effect of the secondary peak in differential permeability, which is observed experimentally in the plastic deformation region, and its dependence on the angle of misalignment between the hard boundary and the soft grain.

Author Contributions: Conceptualization, M.H. and A.K.; methodology, M.H. and A.K.; software, M.H.; validation, M.H. and S.A.; data curation, S.A. and E.H.; writing—original draft preparation, M.H.; writing—review and editing, M.H. and A.K.; visualization, M.H. and A.K.; supervision, A.K. and E.H. All authors have read and agreed to the published version of the manuscript.

Funding: This research received no external funding.

Data Availability Statement: The raw data supporting the conclusions of this article will be made available by the authors on request.

Conflicts of Interest: The authors declare no conflicts of interest.

References

1. Saleem, A.; Goldbaum, D.; Brodusch, N.; Gauvin, R.; Chromik, R.R. Microstructure and mechanical property connections for a punched non-oriented electrical steel lamination. *Mater. Sci. Eng. A* **2018**, *725*, 456–465. [CrossRef]
2. Vourna, P.; Ktena, A.; Tsakiridis, P.E.; Hristoforou, E. A novel approach of accurately evaluating residual stress and microstructure of welded electrical steels. *NDT E Int.* **2015**, *71*, 33–42. [CrossRef]
3. Takahashi, S.; Zhang, L.; Ueda, T. Magnetic hysteresis minor loops in Fe single crystal. *J. Phys. Condens. Matter* **2003**, *15*, 7997–8002. [CrossRef]
4. Iordache, V.E.; Hug, E.; Buiron, N. Magnetic behaviour versus tensile deformation mechanisms in a non-oriented Fe–(3 wt.%)Si steel. *Mater. Sci. Eng. A* **2003**, *359*, 62–74. [CrossRef]
5. Kobayashi, S.; Miura, K.; Narita, Y.; Takahashi, S. Magnetic investigations of steel degradation using a magnetic hysteresis scaling technique. *Metals* **2017**, *8*, 2. [CrossRef]
6. Hristoforou, E.; Vourna, P.; Ktena, A.; Švec, P. On the universality of the dependence of magnetic parameters on residual stresses in steels. *IEEE Trans. Magnet.* **2016**, *52*, 6201106. [CrossRef]
7. Baldev, R.; Jayakumar, T.; Moorthy, V.; Vaidyanathan, S. Characterisation of microstructures, deformation, and fatigue damage in different steels using Magnetic Barkhausen Emission Technique. *Russ. J. Nondestruct. Test.* **2001**, *37*, 789–798. [CrossRef]
8. Franco, F.A.; Padovese, L. NDT flaw mapping of steel surfaces by continuous magnetic Barkhausen noise: Volumetric flaw detection case. *NDT E Int.* **2009**, *42*, 721–728. [CrossRef]
9. Mierczak, L.; Jiles, D.C.; Fantoni, G. A new method for evaluation of mechanical stress using the reciprocal amplitude of magnetic Barkhausen noise. *IEEE Trans. Magn.* **2011**, *47*, 459–465. [CrossRef]
10. Stupakov, O.; Perevertov, O.; Stoyka, V.; Wood, R. Correlation between hysteresis and Barkhausen noise parameters of electrical steels. *IEEE Trans. Magn.* **2010**, *46*, 517–520. [CrossRef]

11. Ktena, A.; Hristoforou, E.; Gerhardt, G.J.; Missell, F.P.; Landgraf, F.J.; Rodrigues, D.L.; Alberteris-Campos, M. Barkhausen noise as a microstructure characterization tool. *Phys. B Condens. Matter* **2014**, *435*, 109–112. [CrossRef]
12. Chmielewski, M.; Piotrowski, L.; Augustyniak, B. A fast procedure of stress state evaluation in magnetically anisotropic steels with the help of a probe with adjustable magnetizing field direction. *Meas. Sci. Technol.* **2017**, *28*, 045903. [CrossRef]
13. Hristoforou, E.; Ktena, A.; Gong, S. Magnetic Sensors: Taxonomy, Applications, and New Trends. *IEEE Trans. Magn.* **2019**, *55*, 4002814. [CrossRef]
14. Angelopoulos, S.; Misiaris, D.; Banis, G.; Liang, K.; Tsarabaris, P.; Ktena, A.; Hristoforou, E. Steel health monitoring device based on Hall sensors. *J. Magn. Magn. Mater.* **2020**, *515*, 167304. [CrossRef]
15. Ktena, A.; Hristoforou, E. Stress Dependent Magnetization and Vector Preisach Modeling in Low Carbon Steels. *IEEE Trans. Magn.* **2012**, *48*, 1433–1436. [CrossRef]
16. Dupré, L.; Sablik, M.J.; Van Keer, R.; Melkebeek, J. Modelling of microstructural effects on magnetic hysteresis properties. *J. Phys. D Appl. Phys.* **2002**, *35*, 2086. [CrossRef]
17. Ducharne, B.; Gupta, B.; Hebrard, Y.; Coudert, J.B. Phenomenological Model of Barkhausen Noise Under Mechanical and Magnetic Excitations. *IEEE Trans. Magn.* **2018**, *54*, 6202606. [CrossRef]
18. Fagan, P.; Ducharne, B.; Daniel, L.; Skarlatos, A. Multiscale modelling of the magnetic Barkhausen noise energy cycles. *J. Magn. Magn. Mater.* **2021**, *517*, 167395. [CrossRef]
19. Sablik, M.J.; Kwun, H.; Burkhardt, G.L.; Jiles, D.C. Model for the effect of tensile and compressive stress on ferromagnetic hysteresis. *J. Appl. Phys.* **1987**, *61*, 3799–3801. [CrossRef]
20. Rekik, M.; Daniel, L.; Hubert, O. Equivalent Stress Model for Magnetic Hysteresis Losses Under Biaxial Loading. *IEEE Trans. Magn.* **2014**, *50*, 2001604. [CrossRef]
21. Ungár, T. Strain Broadening Caused by Dislocations. *Mater. Sci. Forum* **1998**, *278–281*, 151–157. [CrossRef]
22. Li, J.; Xu, M. Modified Jiles-Atherton-Sablik model for asymmetry in magnetomechanical effect under tensile and compressive stress. *J. Appl. Phys.* **2011**, *110*, 063918. [CrossRef]
23. Aydin, U.; Rasilo, P.; Martin, F.; Singh, D.; Daniel, L.; Belahcen, A.; Kouhia, R.; Arkkio, A. Modeling the Effect of Multiaxial Stress on Magnetic Hysteresis of Electrical Steel Sheets: A Comparison. *IEEE Trans. Magn.* **2017**, *53*, 2000904. [CrossRef]
24. Stashkov, A.N.; Kuleev, V.G.; Shchapova, E.A.; Nichipuruk, A.P. Studying Field Dependence of Reversible Magnetic Permeability in Plastically Deformed Low-Carbon Steels. *Russ. J. Nondestruct. Test.* **2018**, *54*, 855–860. [CrossRef]
25. Donahue, M.J.; Porter, D.G. *OOMMF User's Guide, Version 1.0*; NISTIR, no. 6376; National Institute of Standards and Technology: Gaithersburg, MD, USA, 1999.
26. Ktena, A.; Hasicic, M.; Landgraf, F.J.G.; Moudilou, E.; Angelopoulos, S. On the use of differential permeability and magnetic Bark-hausen Noise Measurements for Magnetic NDT Applications. *J. Magn. Magn. Mater.* **2022**, *546*, 168898. [CrossRef]
27. Landgraf, F.J.G.; da Silveira, J.R.F.; Rodrigues, D., Jr. Determining the effect of grain size and maximum induction upon coercive field of electrical steels. *J. Magn. Magn. Mater.* **2011**, *323*, 2335–2339. [CrossRef]

magnetism

Review

Magnetic Hopfions: A Review

Konstantin Guslienko [1,2,3]

1 Departamento de Polimeros y Materiales Avanzados: Fisica, Quimica y Tecnologia, Universidad del Pais Vasco, UPV/EHU, 20018 San Sebastian, Spain; kostyantyn.gusliyenko@ehu.eus
2 EHU Quantum Center, University of the Basque Country, UPV/EHU, 48940 Leioa, Spain
3 IKERBASQUE, the Basque Foundation for Science, 48009 Bilbao, Spain

Abstract: Recent advances in the research area of 3D magnetic topological solitons (hopfions) in restricted geometries are reviewed. The description of the magnetic solitons is based on a macroscopic micromagnetic approach and the Landau–Lifshitz equation of the magnetization motion. The concepts of the gauge emergent vector potential and emergent magnetic field are widely used to calculate the 3D topological charge (the Hopf index) of magnetic textures. The relation of the magnetic hopfions with classical field theory is demonstrated, and a special role of the curvilinear toroidal coordinates in the description of the hopfions is underlined. The hopfion stability and dynamics in ferromagnetic films and dots are considered. A critical discussion of calculations of the magnetization emergent magnetic field and the Hopf index of the toroidal magnetic hopfions in restricted geometries is presented.

Keywords: 3D magnetization textures; topological solitons; Hopf index

1. Introduction

Non-linear configurations of the continuous classical field (solitons) play an important role in the different fields of modern physics such as classical field theory, optics, condensed matter physics, etc. The solitons are localized, particle-like objects, and, in many cases, the solitons can be classified by integer numbers related to their topology (topological charges) [1].

Let us assume that a physical system can be described by a classical 3D vector field $\boldsymbol{n}(\boldsymbol{r})$ of unit length $\boldsymbol{n}^2 = 1$ (the order-parameter space is a unit sphere S_2 in 3D space) depending on three spatial coordinates, represented by the vector $\boldsymbol{r}$. For instance, in the area of ferromagnetism, such a field is the unit magnetization field [2]. The vector field in optics is the Stokes vector field [3,4], whereas in liquid crystals, it is a director field [5,6], etc. There are topologically non-trivial configurations of the vector field $\boldsymbol{n}(\boldsymbol{r})$, which describes an order parameter in the media. Such field configurations can be classified by using a mapping $\boldsymbol{r} \to \boldsymbol{n}(\boldsymbol{r})$ from the coordinate space $(\boldsymbol{r})$ to the order parameter space $\boldsymbol{n}(\boldsymbol{r})$ [7].

The macroscopic equation to describe a magnetization equilibrium configuration and magnetization dynamics in ordered magnetic media (the Landau–Lifshitz equation) is essentially nonlinear and allows for several soliton-type solutions including topological magnetic solitons [2]. Topologically non-trivial magnetization configurations in ferromagnets and ferri- and antiferromagnets, such as domain walls, vortices, skyrmions, and hopfions, etc., are currently a focus of researchers working in the area of solid-state magnetism. However, the existence of the different stable 3D magnetization configurations $\boldsymbol{M}(\boldsymbol{r})$ and the role of 3D (Hopf index) and 2D (skyrmion number) topological charges in the magnetization dynamics are still subject of intensive research. Nowadays, complicated 3D magnetization configurations in ferromagnetic media can be observed experimentally using electron holography or X-ray magnetic imaging [8,9].

The magnetization field $\boldsymbol{M}(\boldsymbol{r})$ is the vector order parameter of a ferromagnet. The magnetization configuration $\boldsymbol{M}(\boldsymbol{r})$ in 3D space is represented by the unit field vector

$\boldsymbol{m}(\boldsymbol{r}) = \boldsymbol{M}(\boldsymbol{r})/|\boldsymbol{M}(\boldsymbol{r})|$ depending on three spatial coordinates $\boldsymbol{r}$ (x_1, x_2, x_3). There are particular cases when the magnetization configurations depend only on one spatial coordinate (domain walls) or two spatial coordinates (magnetic vortices and skyrmions in thin magnetic films or in flat magnetic dots). For stable magnetization configurations $\boldsymbol{m}(\boldsymbol{r})$, it is possible to calculate topological charges, which describe the degrees of mappings (homotopy invariants) of a 1D- (R^1) or 2D coordinate space (R^2) to the unit sphere $\boldsymbol{m}^2 = 1$ in the magnetization space $S_2(\boldsymbol{m})$, i.e., $R^1 \rightarrow S_2(\boldsymbol{m})$, $R^2 \rightarrow S_2(\boldsymbol{m})$. The mapping of the 3D coordinate space $R^3(\boldsymbol{r})$ to the magnetization unit sphere $\boldsymbol{m}^2 = 1$ (S_2) is more complicated and is related to the Hopf index (3D topological charge). The corresponding magnetization configurations with a non-zero Hopf charge are called magnetic hopfions. The 3D magnetization textures are topologically equivalent if they have the same degree of mapping (the Hopf index). The magnetic energy of the given field configuration $E[\boldsymbol{m}(\boldsymbol{r})]$ can be calculated as a functional of the unit magnetization field $\boldsymbol{m}(\boldsymbol{r})$ and spatial derivatives of the vector $\boldsymbol{m}(\boldsymbol{r})$. Such energy includes the exchange, magnetostatic, magnetic anisotropy, etc., energy contributions. The minimization of the energy functional $E[\boldsymbol{m}(\boldsymbol{r})]$ yields some equilibrium (stable or metastable) magnetization field configurations $\boldsymbol{m}(\boldsymbol{r})$. Then, the magnetization dynamics of these equilibrium configurations can be considered on the base of the Landau–Lifshits equation of the magnetization motion, and any parameters, including the topological charges, can be calculated.

In this review, I consider 3D magnetic topological solitons—hopfions and Bloch points. In particular, toroidal hopfions, which were introduced in field theory, are analyzed in detail. Such magnetic solitons reveal intriguing and novel physical properties due to their topologically non-trivial 3D magnetization configurations $\boldsymbol{m}(\boldsymbol{r})$. The article is organized as follows. The basic properties of the hopfions in classical field theory are considered in Section 2. The concept of the emergent electromagnetic field and definition of 3D topological charges of the magnetic textures are presented in Section 3. The static and slow dynamics behaviors of the magnetic hopfions are considered in Section 4. The fast linear and nonlinear hopfion dynamics are considered in Section 5. The review is concluded by a summary in Section 6.

2. Hopfions in the Classical Field Theory

Topological three-dimensional solitons with a non-zero Hopf charge, named hopfions, were first introduced in classical field theory in connection with the non-linear σ-model [7,10–12]. The explicit form of the mapping of a 3D space (R^3) to the surface of the unit sphere $\boldsymbol{n}^2 = 1$ of the classical field $\boldsymbol{n}(\boldsymbol{r})$, $R^3 \rightarrow S_2$, was introduced by Hopf in 1931 [13]. Later, it was found that the Hopf charge (a degree of the mapping $R^3 \rightarrow S_2(\boldsymbol{n})$) can be expressed as some integral of the function composed by a continuous classical field $\boldsymbol{n}(\boldsymbol{r})$ (for instance, the magnetization field $\boldsymbol{m}(\boldsymbol{r})$) and spatial derivatives of the field $\boldsymbol{n}(\boldsymbol{r})$ [14].

Faddeev [15] suggested a new Lagrangian to describe the field $\boldsymbol{n}(\boldsymbol{r})$ (sometimes referred to as the Faddeev's or Faddeev–Skyrme´s Lagrangian), which has stable soliton solutions for the three-component classical field $\boldsymbol{n}(\boldsymbol{r})$ in 3D coordinate space. The stable solutions conserve a 3D topological charge (the Hopf charge). The Faddeev–Skyrme´s Lagrangian is a linear combination of two invariants of the rotation group O(3) represented by the spatial derivatives of the vector field $\boldsymbol{n}(\boldsymbol{r})$. Then, de Vega [16] explicitly found the field $\boldsymbol{n}(\boldsymbol{r})$ components (calling the field configurations "closed vortices") for the simplest non-trivial unit Hopf charge in the toroidal coordinates. Such hopfions are now called toroidal hopfions or torus-like vortex rings. Nicole [17] suggested an analytical form of the Hopf mapping of 3D-coordinate space R^3 to the surface of the unit sphere $\boldsymbol{n}^2 = 1$, $R^3 \rightarrow S_2(\boldsymbol{n})$, which allowed for the explicit introduction of analytic equations for the soliton field configuration with the unit Hopf index. Later, motivated by the excellent paper by Faddeev and Niemi [7], a series of papers on the toroidal hopfions in classical field theory were published [10–12,18]. Gladikowski et al. [10] introduced the field configurations $\boldsymbol{n}(\boldsymbol{r})$ of the toroidal hopfions with an arbitrary Hopf index and presented explicit expressions for the emergent magnetic field and vector potential components in the toroidal coordinates. The toroidal coordinates play a special role in the theory of the hopfions because the analytical equations are essentially simplified in these curvilinear coordinates. The toroidal hopfions attracted considerable attention from researchers because such topological solitons are special stable solutions of the Faddeev–Skyrme´s model [15]. A recent review can be found in Ref. [1].

For calculations of the soliton topological charges, it is important to distinguish localized and non-localized solitons. The field $\boldsymbol{n}(\boldsymbol{r})$ approaches a constant vector value $\boldsymbol{n}_0$ at infinity $|\boldsymbol{r}| \to \infty$ for the localized solitons. The field is inhomogeneous at infinity $|\boldsymbol{r}| \to \infty$ for non-localized solitons. The condition $\boldsymbol{n}(\boldsymbol{r}) \to \boldsymbol{n}_0$ means that the Hopf index, which characterizes the different homotopy classes for the mapping $R^3 \to S_2(\boldsymbol{n})$, is an integer number $\pi_3(S_2) = Z$ in infinite media [7]. Therefore, the toroidal hopfions, which are described by an integer Hopf index $Q_H = 0, \pm1, \pm2 \ldots$, represent a class of the hopfions in ordered media. The Hopf index of the toroidal hopfions can be represented as a product of two winding numbers [10], the planar winding (the azimuthal vorticity) and the twisting of the field configuration $\boldsymbol{n}(\boldsymbol{r})$ along the hopfion tube (the poloidal vorticity), respectively. Due to the property that the vector field is asymptotically trivial, $\boldsymbol{n}(\boldsymbol{r}) \to \boldsymbol{n}_0$ at $|\boldsymbol{r}| \to \infty$, it is possible to compactify the three-dimensional coordinate space to the four-dimensional unit sphere surface, $R^3 \to S_3$. The toroidal hopfions as localized solitons of a classical three-dimensional vector field resemble particle-like objects.

To illustrate the toroidal hopfion field $\boldsymbol{n}(\boldsymbol{r})$, we introduce the toroidal coordinates $\boldsymbol{r}(\eta, \beta, \varphi)$. The connection between the cylindrical (ρ, φ, z) and toroidal (η, β, φ) coordinates is $\rho = a\tau^{-1}sinh(\eta)$, $z = a\tau^{-1}sin(\beta)$, $\varphi = \varphi$, and $\tau = cosh(\eta) - cos(\beta)$, where the toroidal parameter η varies from 0 to ∞, the poloidal angle β varies from $-\pi$ to π, the azimuthal angle φ varies from 0 to 2π, and a is a scale parameter (the hopfion radius in the plane $z = 0$). The toroidal hopfion field components are [19]

$$\begin{aligned} n_z(\eta) &= p\frac{1-cosh^{2m}(\eta)tanh^{2n}(\eta)}{1+cosh^{2m}(\eta)tanh^{2n}(\eta)}, \\ n_x(\boldsymbol{r}) + in_y(\boldsymbol{r}) &= \sqrt{1-n_z^2(\eta)}exp[i(n\varphi + m\beta)], \end{aligned} \tag{1}$$

where $p = n_z(\eta = 0)$ is the hopfion polarity, $p = \pm1$, and the integer numbers m, n are the hopfion poloidal and azimuthal vorticities, respectively.

The profiles of the out-of-plane field component, $n_z(\rho)$, of the toroidal hopfions are shown in Figure 1 for the different hopfion vorticities (m, n). The profiles can be plotted by substituting the poloidal coordinate $\eta(\rho, z) = atanh\left(2a\rho/\left(\rho^2 + z^2 + a^2\right)\right)$ to Equation (1). The component $n_z(0) = n_z(\infty) = 1$ and $n_z(a) = -1$ at the hopfion radius $\rho = a$ in the basal plane $z = 0$ if the hopfion polarity $p = +1$.

Nowadays, different kinds of hopfions are investigated in the field of condensed matter physics (magnetic media [20], liquid crystals and colloids [5], ferroelectrics [21]) and in electromagnetism and gravitation [22] photonics [3], optics [4], etc. There is a growing interest in 3D inhomogeneous magnetization textures classified by a linking number of the preimages of two different points on an $S_2(\boldsymbol{m})$ unit sphere in the 3D coordinate space (R^3), i.e., by the non-zero Hopf index. For a given magnetization texture $\boldsymbol{m}(\boldsymbol{r})$, it is possible to plot preimages, the 3D curves in real space, $\boldsymbol{r}(\boldsymbol{m}_1)$ and $\boldsymbol{r}(\boldsymbol{m}_2)$ of two points $\boldsymbol{m}_1$ and $\boldsymbol{m}_2$ on the magnetization unit sphere $\boldsymbol{m}^2 = 1$ and visually check the number of their crossings (a linking number of the preimages). The linking number introduced solely for the toroidal hopfions is an integer number by definition and is equal to the Hopf index of a magnetization texture $\boldsymbol{m}(\boldsymbol{r})$. However, aside from the toroidal hopfions, there are other kinds of hopfions, and the Hopf index is, in general, not integer and cannot be calculated as the linking number of preimages. Although, 3D localized topological magnetic solitons were introduced by Dzyaloshinskii et al. [23] a long time ago, and the reincarnation of interest to such magnetization field 3D textures started after the publication of the papers by Sutcliffe [24,25] relatively recently.

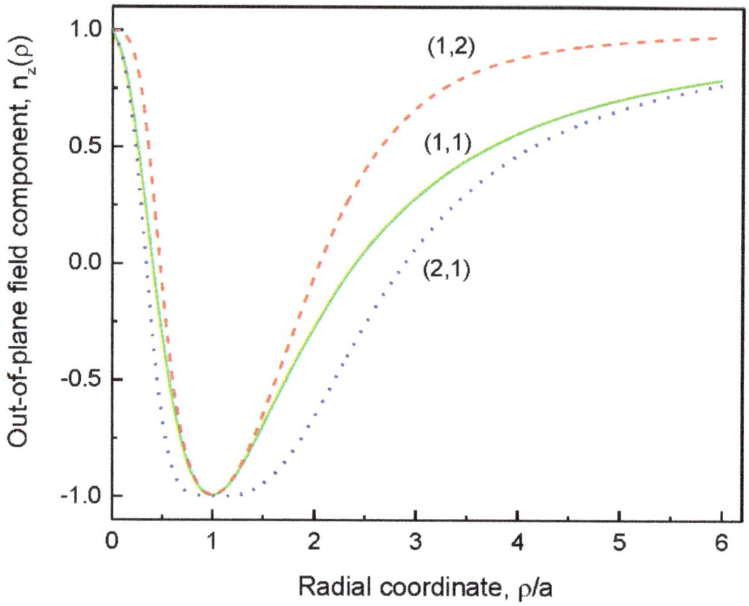

Figure 1. The profiles of the out-of-plane field component, $n_z(\rho)$, of the toroidal hopfions in the plane $z = 0$. The green solid line corresponds to $(m, n) = (1, 1)$, the red dashed line—$(m, n) = (1, 2)$, and the blue dotted line—$(m, n) = (2, 1)$. The hopfion polarity $p = +1$.

The simplest magnetic toroidal hopfions with the Hopf index $|Q_H| = 1$ were considered in infinite ferromagnetic films [26,27] and cylindrical dots [25,28–30] using the vector field toroidal hopfion ansatz [10,11]. It was shown numerically that the toroidal hopfions with $|Q_H| = 1$ can be the ground state of circular ferromagnetic nanodots [29] assuming a strong surface uniaxial magnetic anisotropy. The first experimental observation of the 3D magnetization configurations interpreted as magnetic hopfions was carried out in the Ir/Co/Pt multilayer films [31]. It is shown in Ref. [32] that aside from the toroidal magnetic hopfions, there is, in soft magnetic materials, another class of the magnetization textures with a non-zero Hopf index—the Bloch points. The skyrmion-like winding magnetic hopfions are considered in Section 5.

For the simplest magnetic hopfion with the vorticities $m = 1,\ n = 1,\ |Q_H| = 1$, the hopfion magnetization field components (1) in the cylindrical coordinates are [11]

$$m_z = 1 - \frac{8\rho^2 a^2}{(a^2+\rho^2+z^2)^2},\ m_x + i m_y = \frac{4a\rho exp(i(\varphi+\varphi_0))\left(2za + i(\rho^2+z^2-a^2)\right)}{\left(a^2+\rho^2+z^2\right)^2}. \quad (2)$$

It is evident from Equation (2) that the toroidal hopfion magnetization field $\boldsymbol{m}(\boldsymbol{r})$ approaches a constant vector at infinity $|\boldsymbol{r}| \to \infty$, $\boldsymbol{m}(\infty) = (0, 0, 1)$, as it should be for the localized solitons. The constant angle φ_0 describes a freedom of choosing the directions of the Ox and Oy coordinate axes in the basal xOy plane. According to Ref. [27], we can distinguish Bloch hopfions ($\varphi_0 = 0, \pi$) and Neel hopfions $\varphi_0 = \pm\pi/2$ analogously to the Bloch and Neel skyrmions [33]. The magnetization rotates in the $\hat{\boldsymbol{\varphi}}\hat{\boldsymbol{z}}$ plane ($m_\rho = 0$) for the Bloch hopfions or in the $\hat{\boldsymbol{\rho}}\hat{\boldsymbol{z}}$ plane ($m_\varphi = 0$) for the Neel hopfions when one goes from the coordinate system origin $\boldsymbol{r} = 0$ to the hopfion radius $\rho = a$ along the radial direction $\hat{\boldsymbol{\rho}}$ in the xOy plane.

The in-plane magnetization configurations ($z = 0$) of the toroidal Bloch and Neel hopfions plotted according to Equation (2) are shown in Figure 2. The spatial distribution of the magnetization components m_x, m_z of the toroidal Bloch hopfion ($\varphi_0 = 0$) with the vorticities $(m, n) = (1, 1)$ plotted in the xz plane is shown in Figure 3.

A

B

Figure 2. The in-plane magnetization components m_x, m_y of the toroidal hopfions with the vorticities $(m, n) = (1, 1)$ in the $z = 0$ plane calculated by Equation (2): (**A**) the Bloch hopfion, $\varphi_0 = 0$; (**B**) the Neel hopfion, $\varphi_0 = -\pi/2$.

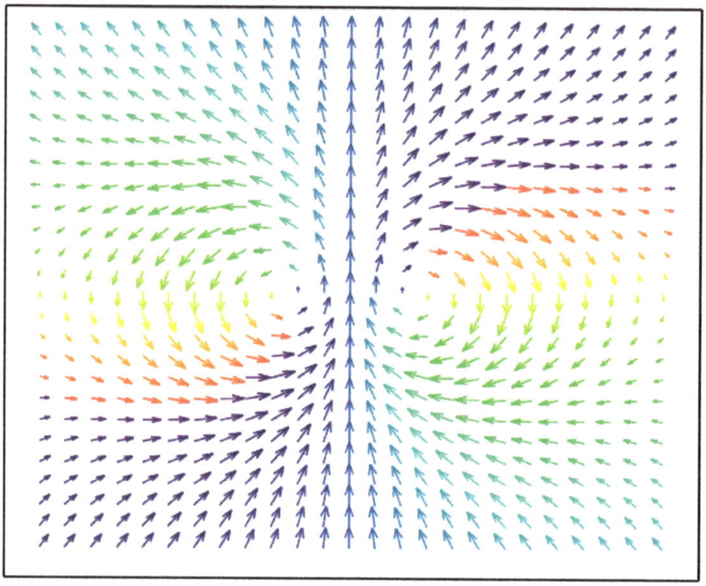

Figure 3. The magnetization components m_x, m_z of the toroidal Bloch hopfion ($\varphi_0 = 0$) with the vorticities $(m, n) = (1, 1)$ in the xz plane calculated by Equation (2).

3. Emergent Electromagnetic Field and the Hopf Index

An inhomogeneous and moving magnetization texture $\boldsymbol{m}(\boldsymbol{r}, t)$ results in the appearance of effective magnetic and electric fields, which act on other subsystems. These so-called emergent magnetic and electric fields are related to the spatial and time derivatives of the magnetization field $\boldsymbol{m}(\boldsymbol{r}, t)$. Initially, this effect was calculated by Korenman et al. [34] and Volovik [35] as a result of the exchange interaction of the local spins with the spins of itinerant electrons in ferromagnetic metals. Then, it was shown [36] that this is a more general, pure geometrical effect related to the choice of the local moving coordinate frame with the Oz axis parallel to the local instant magnetization $\boldsymbol{m}(\boldsymbol{r}, t)$.

The emergent electromagnetic field tensor can be written as

$$F_{\mu\nu}(\boldsymbol{m}) = \boldsymbol{m}\cdot(\partial_\mu \boldsymbol{m} \times \partial_\nu \boldsymbol{m}), \tag{3}$$

where $\partial_\mu = \partial/\partial x_\mu$ denotes spatial and time derivatives. The indices $\mu,\ \nu = 0,\ 1,\ 2, 3$, where $x_0 = ct$ and $\boldsymbol{r} = (x_1, x_2, x_3)$, correspond to the components of a 3D radius–vector $\boldsymbol{r}$ in an orthogonal coordinate system. The field tensor (represented in the units of $\Phi_0/4\pi$, where $\Phi_0 = h/e$ is the magnetic flux quantum) is related to the emergent field four-component vector potential $\boldsymbol{\mathcal{A}} = (A_0, A_1, A_2, A_3)$ as $F_{\mu\nu}(\boldsymbol{m}) = \partial_\mu \mathcal{A}_\nu(\boldsymbol{m}) - \partial_\nu \mathcal{A}_\mu(\boldsymbol{m})$. The emergent magnetic field (sometimes called gyrocoupling density) $\boldsymbol{B} = \nabla \times \boldsymbol{A}$, $\boldsymbol{A} = (A_1, A_2, A_3)$ can be defined as in classical electrodynamics $B_i(\boldsymbol{m}) = \varepsilon_{ijk}F_{jk}(\boldsymbol{m})/2$ [36], where the indices i, j, k mark the spatial coordinates x_i. It is important that the emergent field $\boldsymbol{B}$ is a divergence-free vector field, $\nabla\cdot\boldsymbol{B} = 0$. The flux of the emergent magnetic field $\boldsymbol{B} = \nabla \times \boldsymbol{A}$ through a closed surface defines a 2D topological charge (skyrmion number).

The emergent magnetic field introduced above differs from the real magnetic field defined in standard electrodynamics and is "fictious" in some sense. Nevertheless, the emergent field results in some experimentally measurable effects. Prominent examples of such effects are the topological Hall effect (influence of the emergent magnetic field on trajectories of the conductivity electrons) and skyrmion Hall effect (appearance of the magnetic soliton velocity component transverse to the driving force direction due to the gyroforce). A theoretical approach to exploiting the topological Hall resistance to electrically detect the magnetic hopfion 3D magnetization textures was recently suggested by Göbel et. al. [30]. The flux of the emergent magnetic field trough the plane z = const in 2D ferromagnets determines the gyrovector, an important parameter to describe the motion of 2D topological solitons, magnetic vortices, and skyrmions. The consequences of the non-zero gyrovector in 2D nanostructures have been investigated many times experimentally. A prominent example is the vortex/skyrmion gyrotropic excitation mode immediately related to the gyrovector (see a recent review [33]).

The dot product $\boldsymbol{A}\cdot\boldsymbol{B}$ of the emergent field vector potential $\boldsymbol{A}$ and the emergent magnetic field $\boldsymbol{B}$ defines a 3D topological charge or the Hopf index [37]

$$Q_H = \frac{1}{(4\pi)^2}\int d^3\boldsymbol{r}\boldsymbol{A}(\boldsymbol{m})\cdot\boldsymbol{B}(\boldsymbol{m}) \tag{4}$$

of a 3D magnetization texture $\boldsymbol{m}(\boldsymbol{r})$. There is a gauge freedom in choosing the vector potential $\boldsymbol{A}$. To have a physical sense, the Hopf index should be gauge invariant. The necessary condition for infinite samples is either the nullification of the emergent magnetic field at the sample borders S, $\boldsymbol{B}_S = 0$, or the field $\boldsymbol{B}$ being tangential to the sample surface, $\boldsymbol{B}\cdot\boldsymbol{\sigma} = 0$, where $\boldsymbol{\sigma}$ is the external normal to the surface. The condition $\boldsymbol{B}_S = 0$ is satisfied for the localized solitons (toroidal hopfions, for instance) in infinite samples. The condition $\boldsymbol{B}\cdot\boldsymbol{\sigma} = 0$, in general, is not satisfied at the sample surface. It was shown in Ref. [19] for the toroidal hopfions that only a particular choice of $\boldsymbol{A} = -2\nabla\gamma + \boldsymbol{A}^e$, $\boldsymbol{A}^e = (1 - cos\Theta)\nabla\Phi$, related to the hopfion helicity γ leads to the integer and invariant values of the Hopf indices Q_H in infinite samples, allowing for considering them as the degrees of mapping (the linking numbers or integer numbers of crossings of the magnetization configuration $\boldsymbol{m}(\boldsymbol{r})$ preimages). Here, $\Theta(\boldsymbol{r}) = \Theta(\rho, z)$, $\Phi(\boldsymbol{r}) = n\varphi + \gamma(\rho, z)$, are the spherical angles Θ, Φ of the magnetization, and the hopfion helicity is represented by the poloidal angle β as $\gamma(\rho, z) = m\beta(\rho, z)$, $\beta(\rho, z) = atan\left(2az/\left(\rho^2 + z^2 - a^2\right)\right)$ [19]. The gauge invariance should be considered with respect to this particular choice of the vector potential $\boldsymbol{A} = -2\nabla\gamma + \boldsymbol{A}^e$. In the case of finite samples and magnetization textures different from the toroidal hopfions, the situation is more complicated. It is difficult to choose a proper form of the emergent field vector potential and prove the gauge invariance of the Hopf index. The important question about an integer Hopf index and its gauge invariance should be carefully investigated for each magnetization texture in a restricted geometry.

Recently, Zheng et al. [38] reported on a direct experimental observation of the magnetic toroidal hopfions forming coupled states with skyrmion tube strings in the submicron FeGe plates. They also provided a theoretical interpretation of the observed hopfions (hopfion rings). Zheng et al. [38] used the hopfion ring magnetization to calculate the Hopf invariant (Hopf index) by applying the concept of the emergent magnetic field. The hopfion topological charge (Q_H) was calculated for confined samples. However, the large values of $Q_H = 5,\ 6,\ 10,\ 12$ are very strange from the point of view of the theory of 3D magnetization configurations (including the magnetic hopfions [7,19,27]). The calculation method resulted in such large integer values of Q_H, and gauge invariance of the calculated values of the Hopf index Q_H should be carefully analyzed.

Obviously, the magnetization configurations considered by Zheng et al. are not the so-called toroidal hopfions (torus-like vortex rings) introduced in field theory for infinite media; see Refs. [19,27] and references therein. It was proved that the Hopf index is an integer only for the toroidal hopfions in infinite space (see Section 2). In [38], Zheng et al. considered neither toroidal magnetic hopfions nor infinite samples. Therefore, there is no ground for the speculations about the integer Hopf index and its equivalence to the linking number of their 3D magnetization textures. The Hopf index can be calculated by Equation (4) and is, in general, non-integer for arbitrary 3D magnetization configurations in a finite domain like the ones considered in the paper by Zheng et al. [38].

The strong statement about the integer Hopf index of an arbitrary magnetization configuration in a finite sample was made by Zheng et al. without proof or a proper reference. In modern hopfion theory, such a statement is only valid for the toroidal hopfions in infinite media. The simulations of the equilibrium magnetization configurations $\boldsymbol{m}(\boldsymbol{r})$ and the corresponding Hopf indices were conducted by Zheng et al. for finite samples: 0.5 μm diameter circular dot with a thickness of 180 nm; the square plate, 180 × 1000 × 1000 nm or "bulk" sample 700 × 350 × 350 nm. None of the samples used for the simulations in Ref. [38] is an infinite sample.

The Hopf index can be calculated only if an explicit expression for the vector potential **A** of the emergent magnetic field is known. The authors of Ref. [38] suggested a heuristic equation to calculate the vector potential $\boldsymbol{A}$. However, they did not discuss the important problem of the vector potential gauge and the gauge invariance of the Hopf index. It was shown in Ref. [27] that to make the Hopf index (4) gauge invariant, the emergent field $\boldsymbol{B}(\mathbf{r})$ components should decay fast enough when approaching zero and increasing the absolute value of the radius vector **r** up to infinity $|\boldsymbol{r}| \to \infty$. This is possible to implement only in an infinite sample. In other words, the Hopf index is invariant and does not depend on the vector potential gauge in infinite samples. Even if one introduces the artificial boundary condition for the magnetization $\boldsymbol{m}$ in a finite sample at the sample borders ($\boldsymbol{m} = \boldsymbol{m}_0 = const$) as Zheng et al. [38] suggested, such boundary conditions do not guarantee that the emergent magnetic field $\boldsymbol{B}$ goes to zero at the borders because this field is defined via the magnetization space derivatives, not via the magnetization $\boldsymbol{m}$ itself. There is a specific case of the axially symmetric magnetization configuration $\boldsymbol{m}(\boldsymbol{r}) = \boldsymbol{m}(\rho, z)$ in a cylindrical sample [19,27] ($\boldsymbol{r} = (\rho, \varphi, z)$ are the cylindrical coordinates), for which $\boldsymbol{B} = 0$ at the sample surface if $\boldsymbol{m} = \boldsymbol{m}_0 = \hat{\boldsymbol{z}}$ ($\hat{\boldsymbol{z}}$ is a unit vector in the out-of-plane direction) at the borders. None of the magnetization configurations simulated by Zheng et al. is axially symmetric. Moreover, in many cases, the square plates or "bulk" rectangular samples were used for the simulations. The boundary condition $\boldsymbol{m} = \hat{\boldsymbol{z}}$ used in Ref. [38] means a strong surface uniaxial magnetic anisotropy. It has no physical sense for such magnetic material as FeGe with a cubical crystal structure [39]. Therefore, we can exclude this particular case of axial symmetry from the consideration. It is important that if $\boldsymbol{B} \neq 0$ at the sample borders, then the Hopf index is not gauge invariant [27] and cannot be used to characterize a magnetization texture $\boldsymbol{m}(\boldsymbol{r})$. Even if one can suggest an artificial expression for the vector potential like the equation used by Zheng et al. to satisfy the equation $\boldsymbol{B} = \nabla \times \boldsymbol{A}$, this does not mean that the introduced vector potential is correct.

An expression, alternative to Equation (4) for the Hopf index, is suggested in Refs. [40,41]. This expression is based on the double volume integral from the emergent magnetic field $\boldsymbol{B}(\mathbf{r})$ components and can be derived as a result of the application of the Helmholtz theorem [42] (the fundamental theorem of vector calculus) to the field $\boldsymbol{B}(\mathbf{r})$. An important consequence of the Helmholtz theorem used in Refs. [40,41] is the integral representation of the vector potential $\boldsymbol{A}$ via the emergent field $\boldsymbol{B}$, namely

$$\boldsymbol{A}(\boldsymbol{r}) = -\frac{1}{4\pi}\int d^3\boldsymbol{r}' \frac{\boldsymbol{R} \times \boldsymbol{B}(\boldsymbol{r}')}{R^3}, \quad \boldsymbol{R} = \boldsymbol{r} - \boldsymbol{r}'. \tag{5}$$

This expression for $\boldsymbol{A}(\boldsymbol{r})$ (5) is different from the expression for the emergent field vector potential given by Zheng et al. [38]. Therefore, the heuristic vector potential introduced by Zheng et al. contradicts the Helmholtz theorem [42] and the integral expressions for the Hopf index used in Refs. [40,41]. The similar wrong heuristic expressions for the vector potential (represented as one-fold integral from the emergent magnetic field $\boldsymbol{B}$) were used in Refs. [20,43] without any justification.

An expression for the vector potential was used by Liu et al. [28] for calculations of the Hopf index of the circular magnetic dots in the momentum representation. Liu et al. calculated the Hopf index to be 0.96 for their dot parameters. However, Liu et al. [26,28] used the Coulomb gauge for the emergent field vector potential, $\nabla \cdot \boldsymbol{A} = 0$, which is not compatible with Equation (5) for the potential $\boldsymbol{A}(\boldsymbol{r})$ given above. The gauge $\nabla \cdot \boldsymbol{A} = 0$ is unphysical for the particular case of the toroidal hopfions. This can be checked if one uses the toroidal vector potential components calculated explicitly in Ref. [19].

The value of the Hopf index represented by Equation (4) depends essentially on the choice of the vector potential $\boldsymbol{A}$ of the emergent magnetic field. Recently, it was shown [19] that a definite gauge of the vector potential $\boldsymbol{A}$ should be chosen for the toroidal hopfions in infinite samples to guarantee the integer values of the Hopf index Q_H. The heuristic vector potential given by Zheng et al. [38] can guarantee neither the integer Hopf indices nor their stability with respect to the vector potential gauge change in the finite samples. It seems that the integer values of Q_H simulated in the paper by Zheng et al. [38] are some arbitrary and ungrounded numbers resulting in a confusion of the readers who are not specialists in the theory of 3D magnetic textures (hopfions, in particular). Therefore, the methods and calculations used in Ref. [38] should be reconsidered to bring them in accordance with the hopfion theory.

4. Static Properties of Magnetic Hopfions

Recently, the stability of the toroidal hopfions was studied numerically in chiral ferromagnets FeGe [28,29] and in confined ferroelectric nanoparticles (nanospheres of PbZrTi oxide) [21]. Based on the previous theoretical predictions [28,29], the first experimental observation of magnetic toroidal hopfions was carried out in the Ir/Co/Pt multilayer systems [31] with ultrathin Co layers. A restricted sample cylindrical geometry was used for the chiral ferromagnets FeGe and Ir/Co/Pt with a strong Dzyaloshinskyi-Moriya interaction and out-of-plane uniaxial magnetic anisotropy on the sample faces. The calculated Hopf index Q_H for chiral ferromagnets was equal to 1 [29,31] or close to 1 [28]. It was simulated in Ref. [28] that the toroidal hopfions are metastable states of a chiral nanodisk existing at a large-enough disk thickness and radius. Their magnetic energy $E[\boldsymbol{m}(\boldsymbol{r})]$ is higher than the energy of the monopole–antimonopole pair 3D magnetization configuration.

The hopfions were also considered in ferroelectrics. The Hopf index of the polarization field $\boldsymbol{P}$ was defined using a non-standard equation in Ref. [21]. If one assumes that the polarization field $\boldsymbol{P}(\boldsymbol{r})$ for ferroelectrics is similar to the magnetization field $\boldsymbol{M}(\boldsymbol{r})$ for ferromagnets, then the definition of the Hopf index should be the emergent magnetic field $\boldsymbol{B}(\boldsymbol{r})$, not the field $\boldsymbol{M}(\boldsymbol{r})$ (or $\boldsymbol{P}(\boldsymbol{r})$). Therefore, the parameter H defined in Ref. [21] as a volume integral from the dot product $\boldsymbol{P}\cdot\boldsymbol{A}$ is not a Hopf index, although the spatial configuration of the field $\boldsymbol{P}(\boldsymbol{r})$ is similar to the one for the toroidal hopfion of the magnetization field $\boldsymbol{M}(\boldsymbol{r})$. It is not a surprise that the Hopf index H is not an integer for the considered spherical samples. The calculated values of the Hopf index are not listed in Ref. [21].

The magnetization configurations of the toroidal magnetic hopfions in infinite media with the integer Hopf index $Q_H = mn$ and the arbitrary poloidal and azimuthal hopfion vorticities m, n were explicitly calculated in Ref. [19] using the toroidal and cylindrical coordinates. The calculation method was based on the Hopf mapping's explicit definition and the concept of the emergent magnetic field defined via spatial derivatives of the magnetization field $\boldsymbol{m}(\boldsymbol{r})$. It was shown that the Hopf index density can be represented as a Jacobian of the transformation from the toroidal to the cylindrical coordinates [19]. The calculated components of the emergent magnetic field and emergent field vector potential can be used, in particular, for explanations of the topological and skyrmion Hall effects of the toroidal magnetic hopfions.

The separate problem is the stability of the different 3D magnetization textures $\boldsymbol{m}(\boldsymbol{r})$. It was proven within field theory [44,45] that for any physical systems with a squared gradient field term in the Lagrangian, there are no stable, stationary, localized solutions in the 3D case for any form of the potential. This statement is known as the Hobart–Derrick theorem. However, stable localized solutions (localized solitons) may exist if any energy contributions linear with respect to spatial derivatives or with higher-order spatial derivatives of the magnetization field are present in the Lagrangian [46,47]. In the theory of magnetism, there are some specific energy terms with the first-order derivatives, so-called Lifshitz invariants, accounting for the Dzyaloshinskii–Moriya interactions (DMI) in ferro-magnetic materials with broken inversion symmetry. Another opportunity to obtain a stable 3D magnetization field configurations is accounting for the higher-order spatial derivatives in the Lagrangian. The term quartic in spatial derivatives was introduced by Skyrme [48] and Faddeev [15] a long time ago within classical field theory. It was shown that the Faddeev–Skyrme Lagrangian

$$\mathcal{L} = \frac{1}{2}\int d^3\boldsymbol{r}\Big[\left(\partial_\mu \boldsymbol{n}\right)^2 + g\left(F_{\mu\nu}(\boldsymbol{n})\right)^2\Big] \tag{6}$$

has stable 3D localized soliton field solutions in the form of toroidal hopfions [7,10–12]. Here, g is the coupling constant.

The toroidal hopfions belong to a class of the localized topological solitons and are characterized by integer values of a 3D topological charge (Hopf index) [13,14]. It was shown recently [20] that the classical Heisenberg model with competing long-range exchange interactions results in quadratic terms in the second spatial derivatives of the magnetization field. Although such Heisenberg model is beyond the standard theory of micromagnetism, it may lead to the stabilization of the toroidal magnetic hopfions. The question is this: is it possible to stabilize the toroidal magnetic hopfions in a ferromagnet with the standard micromagnetic exchange energy (avoiding exotic exchange interactions) due to non-zero DMI terms and/or magnetostatic energy? Such micromagnetic energy contributions are beyond the field theory frameworks and, therefore, the applicability of the Hobart–Derrick theory to the evaluation of the stability of the magnetization field configurations should be reconsidered. The simple scaling analysis accounting for the DMI energy terms was conducted in Ref. [49]. However, this scaling analysis ignored the finite sample sizes and the magnetostatic interaction (which is unavoidable in real ferromagnetic samples). The complicated non-local magnetostatic energy is usually not accounted for in the theory of magnetic skyrmions and hopfions or accounted for in a simplified form. The skyrmions are usually considered in the bulk ferromagnetic crystals without inversion symmetry or in ultrathin magnetic films. In both cases, the magnetostatic energy is reduced to a local form and accounted for as an extra contribution to the magnetic anisotropy energy. Accounting for the magnetostatic energy in relatively thick ferromagnetic dots [50] allows us to stabilize quasi-2D magnetic skyrmions without the presence of any DMI terms if a small out-of-plane magnetic uniaxial anisotropy is included in the energy functional. The magnetostatic energy was not included in the energy functional in Refs. [25,29,30] describing the toroidal magnetic hopfions in thick cylindrical dots. The hopfion's constant phase angle φ_0 introduced in Section 2 does not contribute to the exchange energy. However, the magnetostatic and DMI energies depend on the particular value of φ_0 and are different for the Bloch and Neel toroidal hopfions. We note that the magnetostatic energy can also lead to the stabilization of other kinds of 3D magnetization textures: the Bloch point hopfions with non-zero Hopf charge or half-hedgehog (3D quasi-Neel skyrmion) magnetization textures even in soft magnetic materials with no DMI [32,51,52].

It is demonstrated in Ref. [53] that the Bloch toroidal hopfion magnetization texture is a metastable state of a thick cylindrical dot or a long cylindrical wire of a finite radius R. The existence of this metastable state is a result of a competition of the exchange and magnetostatic energies. The Dzyaloshinskyi–Moriya exchange energy and magnetic uniaxial anisotropy are of the second importance for the toroidal hopfion stabilization. The important role of the magnetostatic interaction in the toroidal hopfion stabilization in a cylindrical dot was confirmed by simulations [43].

5. Dynamics of Magnetic Hopfions

A numerical study of the spin excitation spectra of the toroidal magnetic hopfions was performed in recent papers [54,55]. Very recently, theoretical papers on the magnetic hopfion dynamics [26,43,56,57] were published.

The spin-polarized electrical current-induced 3D dynamics of the toroidal magnetic hopfions were studied both analytically (the collective coordinates approach) and numerically in Ref. [26] in an infinite frustrated ferromagnet (different signs of the exchange integrals for the nearest neighbor spins). The hopfions exhibit complicated dynamics including a longitudinal motion along the current direction, a transverse motion with respect to the current direction, a rotational motion, and dilation [26]. The skyrmion Hall effect (due to non-zero hopfion gyrovector) is clearly seen in the solutions, although the gyrovector was not explicitly defined. The calculated hopfion velocity is quite small, <10 m/s, for the typical values of the current intensity. The results by Liu et al. [26] are in some contradiction with the hopfion dynamics calculated by Wang et al. in Ref. [27], where the hopfion's internal structure dynamics were ignored. Wang et al. did not observe any skyrmion Hall effect in the course of the toroidal hopfion motion in magnetic stripes. The hopfions moved under the drive of the spin-transfer torque or spin Hall torque along the nanostripe with velocities up to 20 m/s. The origin of this discrepancy is still unclear. The Hopf index was $|Q_H| = 1$ in both cases.

The hopfion dynamics depend essentially on the phase angle φ_0 defining the toroidal hopfion static magnetization (2). The spin wave excitation spectra of the Bloch and Neel toroidal hopfions in FeGe nanodisks were investigated numerically in Ref. [43]. It was shown that the Bloch hopfions reveal very dense eigenfrequencies in the low-frequency region (<5 GHz). Only a few eigenfrequencies were detected for the Neel hopfions. The eigenmode spatial distributions were found for some selected spin excitation modes. The modes were mainly localized at the disk edge for the Bloch hopfions or near the dot center for Neel hopfions [43]. The eigenmode classification according to their symmetry or number of the nodes in different spatial directions were not presented. It seems that the Bloch hopfion frequency spectra are numerical artifacts and further careful analysis is necessary. The Hopf index of the Neel hopfions calculated using the concept of the emergent electromagnetic is above 0.9. It is unclear why the Neel hopfions transform from a circularly symmetric shape (azimuthal symmetry) to a square shape, increasing the axial magnetic field [43]. There is no reason to break the azimuthal symmetry of the static magnetization configurations.

The spin excitation modes of the Bloch toroidal hopfions in nanodisks (the radius R is 100 nm and thickness L is 70 nm) were simulated in Ref. [54] in a wide frequency range of 0–20 GHz. The spin eigenmodes were classified according to their symmetry, number of the nodes in the azimuthal and out-of-plane directions, and degree of the mode localization. The dot edge, dot middle, and dot center localized modes were identified. No simple rule establishing a connection of the mode localization with its frequency was found. The edge and middle localized modes had large azimuthal indices. Such complicated inhomogeneous mode patterns mean that the ferromagnetic resonance intensity (average mode volume magnetization) of the modes is small and it will be difficult to observe these modes by the standard ferromagnetic resonance spectroscopy technique. It was found numerically [55] that the toroidal hopfions have distinctly less resonance peaks in comparison with skyrmion tubes. It was also found in Ref. [55] that the hopfion breathing excitation modes (oscillations of the hopfion radius) and rotating spin modes (magnetization rotates in the hopfion basal plane xOy) hybridize, applying the oscillation magnetic field along the hopfion axis $0z$.

As it is well known, the skyrmion Hall effect (existence of a gyroforce, which is perpendicular to the soliton velocity) for magnetic topological solitons is attributed to a non-zero gyrovector. The gyrovector, in the case of the magnetic vortices and skyrmions, determines their low-frequency dynamics in both 3D and 2D cases [58,59]. Therefore, the problem of the calculation of the magnitude and direction of the gyrovector of the magnetic hopfions is of practical importance. The components of the emergent magnetic field in the cylindrical coordinate system, defining the gyrovector $\mathbf{G}$ of the toroidal magnetic hopfion, are presented in Ref. [27]. It is proven unambiguously [27] that the gyrovector out-of-plane component G_z of an axially symmetric toroidal magnetic hopfion with the Hopf index $|Q_H| = 1$ is equal to zero. Concerning the zero values of the two in-plane gyrovector components, only a plausible assumption is considered that they have to vanish due to the toroidal hopfion and system symmetry. Nevertheless, this assumption was accepted in a number of subsequent articles, where the equality to zero of all components of the toroidal hopfion gyrovector $\mathbf{G}$ was mentioned as a proven fact [56,57,60]. Meanwhile, this issue needs to be clarified, since the possible non-zero values of the gyrovector components affect essentially both the three-dimensional current-induced hopfion translational motion [26] and the spin waves excited over the hopfion background [56] by the external magnetic field. Liu et al. [57] excluded the toroidal hopfion gyrovector from consideration and represented the hopfion dynamics solely in terms of the hopfion emergent magnetic field toroidal $\boldsymbol{T} = 1/2\int dV(\boldsymbol{r}\times\boldsymbol{B})$ and octupole moments. Such an approach resulted in a specific toroidal hopfion dynamic presented in Ref. [57]. The static toroidal hopfion ansatz for the particular case of the Hopf index $|Q_H| = 1$ in Ref. [26] was initially written incorrectly, and then it was corrected in a paper (Ref. [60]) by the same group.

The calculation approach suggested in Ref. [61] is based on the concept of the emergent magnetic field $\boldsymbol{B}(\boldsymbol{r})$ introduced in Section 3 and a calculation of the emergent field components and their volume averages for the toroidal hopfion magnetization texture in the appropriate curvilinear coordinate systems. The simplest non-trivial toroidal hopfion with the Hopf index $|Q_H| = 1$ in the cylindrical magnetic dot was considered, and the dependencies of the Hopfion gyrovector components on the dot sizes were calculated. It was demonstrated by analytical calculations that the magnetic hopfion gyrovector $\boldsymbol{G}$ is not equal to zero and does not vanish even in the limit of an infinite sample. Namely, two components of the gyrovector in the curvilinear cylindrical coordinates, G_z and G_φ, are finite. The out of-plane z-component of the hopfion gyrovector (G_z) goes to zero, increasing the dot radius; however, the in-plane gyrovector φ-component (G_φ) remains finite [61]. The calculated components of the hopfion emergent magnetic field and gyrovector can be used for calculations of the topological and skyrmion Hall effect of the toroidal magnetic hopfions, respectively. It was recently shown [62] that the toroidal hopfions reveal a Hall motion (skyrmion Hall effect) under the current pulses, while the skyrmionium moves only along the current direction. This is because the gyrovector (2D topological charge) of skyrmionium vanishes, whereas it is not equal to zero for the toroidal magnetic hopfions in accordance with Ref. [61]. The fractional Hopf index $Q_H = 0.8$ was calculated in Ref. [62] due to the restricted geometry of FeGe samples. The current pulses were employed to drive the dynamics of the fractional hopfions. An asymmetric Hall motion of the hopfions with respect to the current direction was detected [62].

Space–time magnetic hopfions are suggested in Ref. [63]. They are treated as 2D magnetic textures (skyrmions) excited by an oscillating magnetic or electric field. The authors considered the coupled dynamics of the skyrmion radius and helicity using the model of the exchange interaction quadratic and quartic on the magnetization spatial derivatives. The authors of Ref. [63] believe that the introduced hopfion topological invariant, the space–time Hopf charge ($Q_H = \pm 1$), can be tuned by the applied electric field. The emergent magnetic field $B_i(\boldsymbol{m}) = \varepsilon_{ijk}\boldsymbol{m}\cdot(\partial_j\boldsymbol{m}\times\partial_k\boldsymbol{m})$ $(i, j, k = x, y, t)$ and gauge vector potential Awere introduced using the standard equations. However, these equations are valid only if the indices i, j, k used in their definition [63] mark the spatial coordinates, for instance, the Cartesian coordinates (x, y, z). However, one of the indices i, j, k corresponds to time (t) according to Knapman et al. [63]. If the index $i = t$, then the emergent field component B_t has no physical sense because $\boldsymbol{B}= (B_x, B_y, B_z)$ is a vector in three-dimensional coordinate space (x, y, z). If the indices $j = t$ or $k = t$, then the vector $\boldsymbol{B}$ is not the emergent magnetic field, and the equation $\boldsymbol{B} = \nabla\times\boldsymbol{A}$ is not valid. In the case of $j = t$ or $k = t$, the field $\boldsymbol{B}$ is an emergent electric field $\boldsymbol{E}$ proportional to the time derivative of the unit magnetization vector $\mathbf{m}$. The relation of the emergent electric field defined as $E_i = F_{i0}$ $(i = x, y, z)$ or $\boldsymbol{E} = \nabla A_0 - \partial\boldsymbol{A}/c\partial t$ and the four-vector potential $\boldsymbol{\mathcal{A}} = (A_0, \boldsymbol{A})$ is essentially different from the equation $\boldsymbol{B} = \nabla\times\boldsymbol{A}$ for the emergent magnetic field. Only the space Hopf index defined by Equation (4) via the dot product $(\boldsymbol{A}\cdot\boldsymbol{B})$ has physical sense. The time derivative of the magnetization $\mathbf{m}$ and emergent electric field $\boldsymbol{E}\sim\partial\boldsymbol{m}/\partial t$ cannot be used for the calculation of the Hopf index defined by Equation (4). Therefore, the equations defining the space–time Hopf index used by Knapman et al. [63] are wrong and should be corrected.

The hopfions considered by Knapman et al. are a particular case of 3D winding hopfions introduced by Kobayashi et al. [64] in the $R^2 \times S_1$ coordinate space, where the third coordinate S_1 (z) is directed along the sample thickness, L. Knapman et al. [64] substituted the thickness coordinate (z) to time (t). There is the standard angular parameterization of the magnetization $\boldsymbol{m}(\boldsymbol{r})$ components via the spherical angles $\Theta(\boldsymbol{r})$, $\Phi(\boldsymbol{r})$. The hopfion magnetization spherical angles should be written in an axially symmetric form, $\Theta(\boldsymbol{r}) = \Theta(\rho, z)$, $\Phi(\boldsymbol{r}) = n\varphi + \gamma(\rho, z)$, as it was performed for the azimuthally symmetric toroidal hopfions in Refs. [19,53]. If following Ref. [64] one introduces the hopfion helicity as $\gamma(\rho, z) = -Pg(z)$, where the phase angle $g(z)$ satisfies the boundary condition $g(L) - g(0) = 2\pi$ at the sample upper/down faces and P is an integer, then the calculated Hopf index of the winding hopfion is an integer. Kobayashi et al. assumed that the angle $\Theta(\rho, z)$ does not depend on the thickness coordinate z. Therefore, the dependence $\Theta(\rho)$ describes a 2D skyrmion profile in the xOy plane.

Following Kobayashi et al. [64], we introduce the winding hopfion magnetization components as

$$m_z(\rho) = cos\Theta(\rho),\ m_x(\boldsymbol{r}) + im_y(\boldsymbol{r}) = sin\Theta(\rho)e^{i(n\varphi - Pg(z) + \varphi_0)}, \tag{7}$$

where $\Theta(\rho)$ is the skyrmion magnetization polar angle satisfying the boundary conditions $\Theta(0) = \pi$, $\Theta(\infty) = 0$, n is the azimuthal vorticity, and φ_0 is a constant phase angle.

To plot the winding hopfion magnetization, we take the skyrmion profile in the Belavin–Polyakov form [33] $cos\Theta(\rho) = (\rho^2 - 1)/(\rho^2 + 1)$ (ρ is taken in the units of the skyrmion radius ρ_c, $m_z(\rho_c,) = 0$) and the function $g(z) = 2\pi(z/L)$ in the simplest form satisfying the conditions $g(L) - g(0) = 2\pi$. The magnetization components of the winding hopfions are plotted in Figure 4. We note that the skyrmions at the film surfaces $z = 0$, L are of the Neel type at $\varphi_0 = 0$ (Figure 4A) or Bloch type at $\varphi_0 = \pi/2$ (Figure 4B).

The calculation by Equation (4) yields $Q_H = QP$, where $Q = n$ is the skyrmion number [36] (2D topological charge). However, in the finite samples, Q is not an integer and the condition $g(L) - g(0) = 2\pi$, in general, is not satisfied. Therefore, the Hopf index of the winding hopfions can be an arbitrary number. The winding hopfions are very different from the toroidal magnetic hopfions, which have almost integer Hopf indices even in finite samples [28,29].

Very recently, the paper by Saji et al. [65] was published, where the authors rewrote the toroidal hopfion magnetization in the spherical coordinates regardless of whether the hopfion has a cylindrical (azimuthal) symmetry. The validity of the ansatz derived in Ref. [65] for the Hopf index $Q_H > 1$ should be proven. It is also unclear why the emergent vector potential calculated by Saji et al. [65] differs from the vector potential presented in Ref. [59] calculated on the base of field theory [10,11]. In Refs. [26,60,65], the toroidal hopfion magnetization, emergent magnetic field, and vector potential are presented via an arbitrary smooth function $f(r)$ satisfying the boundary conditions $f(0) = 0$, $f(\infty) = \pi$ to ensure an integer Hopf index Q_H in an infinite sample. Therefore, these important local parameters $\boldsymbol{m}(\boldsymbol{r}), \boldsymbol{B}(\boldsymbol{r}), \boldsymbol{A}(\boldsymbol{r})$ are not completely defined and cannot be used for calculations of any effects related to the toroidal magnetic hopfions. Using the function $f(r)$ is not necessary because the exact explicit expressions for $\boldsymbol{m}(\boldsymbol{r}), \boldsymbol{B}(\boldsymbol{r}), \boldsymbol{A}(\boldsymbol{r})$ of the toroidal magnetic hopfions in infinite media were found in Ref. [19] for any values of the 3D topological charge Q_H.

The authors of Ref. [65] demonstrated that the spin waves excited over the hopfion background experience the emergent electromagnetic field generated by the toroidal hopfion magnetization texture $\boldsymbol{m}(\boldsymbol{r})$. In particular, it was shown that the spin waves propagating along the toroidal hopfion symmetry axis (0z) are deflected by the hopfion magnetic texture. This effect using an analogy with the skyrmion topological Hall effect was named as "magnonic Hall effect".

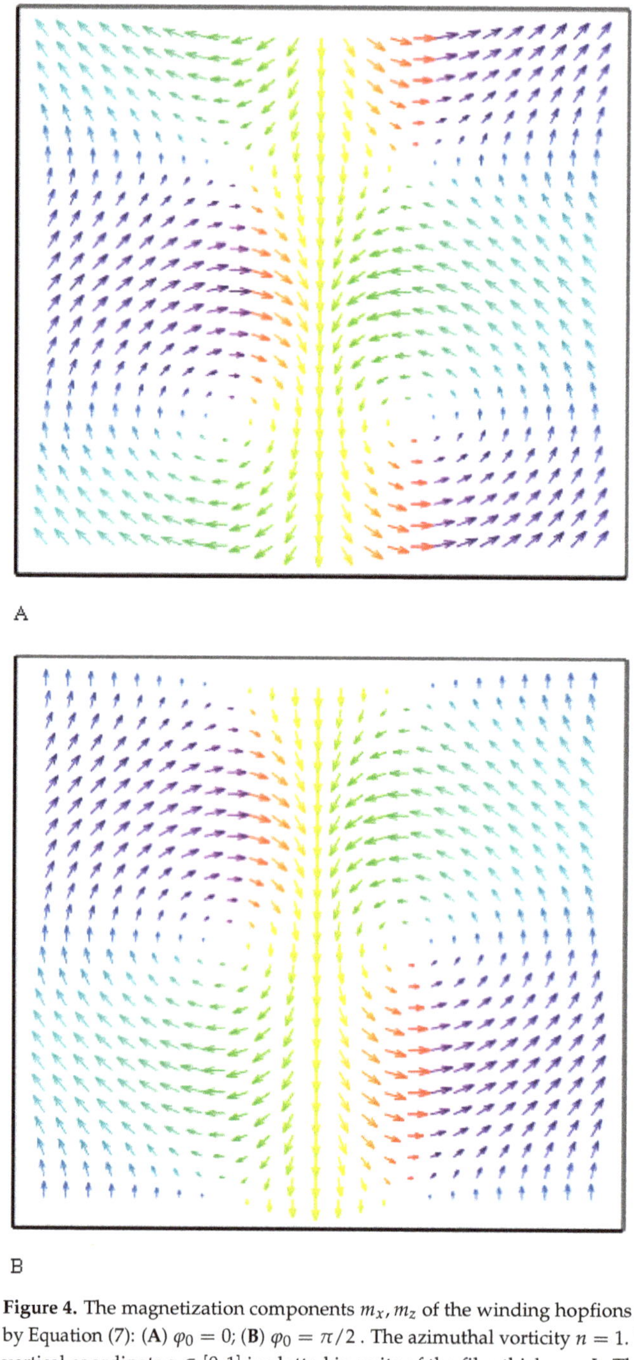

A

B

Figure 4. The magnetization components m_x, m_z of the winding hopfions in the xz plane calculated by Equation (7): (**A**) $\varphi_0 = 0$; (**B**) $\varphi_0 = \pi/2$. The azimuthal vorticity $n = 1$. The parameter $P = 1$. The vertical coordinate $z \in [0, 1]$ is plotted in units of the film thickness L. The horizontal coordinate x varies from -3 to 3 in units of the skyrmion radius ρ_c.

6. Summary

I briefly review above the current status of research on 3D magnetic topological solitons (hopfions) in restricted and infinite geometries. The hopfions are initially introduced, similarly to the skyrmions, in classical field theory. Then, these complicated 3D textures are calculated and measured in ferromagnetic media. It is shown that the concepts of the gauge emergent vector potential and emergent magnetic field are absolutely necessary to describe the magnetic hopfions. The toroidal hopfions with the integer Hopf indices found in the field theory occupy a special place among the magnetic hopfions. However, one should be careful using the equations developed for infinite media to calculate the hopfions in finite samples. Except for the toroidal hopfions, other hopfions with a non-zero and non-integer Hopf index are also possible (the Bloch points and winding hopfions, for instance). I note that the Hopf index (3D topological charge) can be used as an important parameter (invariant) to characterize a given 3D magnetization texture. The magnetization textures with the same values of the Hopf index are topologically equivalent. The Hopf index has less physical sense in comparison to a 2D topological charge (skyrmion number) because the latter defines the soliton gyrovector resulting, in particular, in the vortex/skyrmion gyrotropic excitation mode observed experimentally.

The main articles on the toroidal hopfion stability and dynamics in ferromagnetic films and dots are reviewed. Both topics are considered mainly theoretically (numerically) because it is difficult to observe stable hopfions and their dynamics experimentally. No applications of the magnetic hopfions have been realized to the moment. I believe that it can be achieved in the near future.

A critical discussion on the calculations and simulations of the magnetization, emergent magnetic field, and Hopf index of the toroidal magnetic hopfions and other kinds of magnetic hopfions in restricted geometries is presented. The critical point is using proper equations for the hopfion emergent magnetic field and vector potential and proof of gauge invariance of the calculated Hopf index.

Funding: The author acknowledges support by IKERBASQUE (the Basque Foundation for Science). The research was funded in part by the Spanish Ministry of Science, Innovation and Universities grant PID2022-137567NB-C21/AEI/10.13039/501100011033, by the Basque Country government under the scheme "Ayuda a Grupos Consolidados" (Ref. IT1670-22), and by the National Science Centre of Poland, project no UMO-2023/49/B/ST3/02920.

Institutional Review Board Statement: Not applicable.

Informed Consent Statement: Not applicable.

Data Availability Statement: The data supporting reported results can be found in the preprints by K.Y. Guslienko, "Emergent magnetic field and vector potential of the toroidal magnetic hopfions", 6 February 2024, DOI: 10.21203/rs.3.rs-3930623/v1, and "3D Magnetization Textures: Toroidal Magnetic Hopfion Stability in Cylindrical Samples", 12 December 2023, DOI: 10.20944/preprints202312.0816.v1.

Acknowledgments: The author thanks O. Chubykalo-Fesenko, P. Fischer, A. Hoffmann, G.N. Kakazei, N. Kiselev, R. Knapman, M. Krawczyk, A. Lyberatos, R. Rama-Eiroa, F. Rybakov, I. Smalyukh, E.V. Tartakovskaya, G. Volovik, and J. Zang for stimulating discussions on the magnetic hopfions.

Conflicts of Interest: The author declares no conflicts of interest. The funders had no role in the design of the study; in the collection, analyses, or interpretation of data; in the writing of the manuscript; or in the decision to publish the results.

References

1. Shnir, Y.M. Chapter 8: Hopfions. In *Topological and Non-Topological Solitons in Scalar Field Theories*; Cambridge University Press: Cambridge, UK, 2018; pp. 221–246.
2. Kosevich, A.M.; Ivanov, B.A.; Kovalev, A.S. Magnetic solitons. *Phys. Rep.* **1990**, *194*, 117–238. [CrossRef]
3. Shen, Y.; Yu, B.; Wu, H.; Li, C.; Zhu, Z.; Zayats, A.V. Topological transformation and free-space transport of photonic hopfions. *Adv. Photonics* **2023**, *5*, 015001. [CrossRef]

4. Sugic, D.; Droop, R.; Otte, E.; Ehrmanntraut, D.; Nori, F.; Ruostekoski, J.; Denz, C.; Dennis, M.R. Particle-like topologies in light. *Nat. Commun.* **2021**, *12*, 6785. [CrossRef] [PubMed]
5. Smalyukh, I.I. Review: Knots and other new topological defects in liquid crystals and colloids. *Rep. Prog. Phys.* **2020**, *83*, 106601. [CrossRef]
6. Tai, J.S.B.; Ackerman, P.J.; Smalyukh, I.I. Topological transformations of Hopf solitons in chiral ferromagnets and liquid crystals. *Proc. Natl. Acad. Sci. USA* **2018**, *15*, 921–926. [CrossRef]
7. Faddeev, L.; Niemi, A. Stable knot-like structures in classical field theory. *Nature* **1997**, *387*, 58–61. [CrossRef]
8. Donnelly, C.; Scagnoli, V. Imaging three-dimensional magnetic systems with X-rays. *J. Phys. Cond. Matter* **2020**, *32*, 213001. [CrossRef]
9. Donnelly, C.; Metlov, K.L.; Scagnoli, V.; Guizar-Sicairos, M.; Holler, M.; Bingham, N.S.; Raabe, J.; Heyderman, L.J.; Cooper, N.R.; Gliga, S. Experimental observation of vortex rings in a bulk magnet. *Nat. Phys.* **2021**, *17*, 316–321. [CrossRef]
10. Gladikowski, J.; Hellmund, M. Static solitons with non-zero Hopf number. *Phys. Rev. D* **1997**, *56*, 5194–5199. [CrossRef]
11. Hietarinta, J.; Salo, P. Faddeev-Hopf knots: Dynamics of linked un-knots. *Phys. Lett. B* **1999**, *451*, 60–67. [CrossRef]
12. Hietarinta, J.; Salo, P. Ground state in the Faddeev-Skyrme model. *Phys. Rev. D* **2000**, *62*, 081701. [CrossRef]
13. Hopf, H. Über die Abbildungen der dreidimensionalen Sphäre auf die Kugelfläche. *Math. Analen* **1931**, *104*, 637–665. [CrossRef]
14. Whitehead, J.H.C. An expression of Hopf 's invariant as an integral. *Proc. Natl. Acad. Sci. USA* **1947**, *33*, 117–123. [CrossRef] [PubMed]
15. Faddeev, L.D. Some comments on the many-dimensional solitons. *Lett. Math. Phys.* **1976**, *1*, 289–293. [CrossRef]
16. de Vega, H.J. Closed vortices and the Hopf index in classical field theory. *Phys. Rev. D* **1978**, *18*, 2945. [CrossRef]
17. Nicole, D.A. Solitons with non-vanishing Hopf index. *J. Phys. G Nucl. Phys.* **1978**, *4*, 1363–1369. [CrossRef]
18. Sutcliffe, P. Knots in the Skyrme–Faddeev model. *Proc. R. Soc. A* **2007**, *463*, 3001–3020. [CrossRef]
19. Guslienko, K.Y. Emergent magnetic field and vector potential of the toroidal magnetic hopfions. *Chaos Solitons Fractals* **2023**, *174*, 113840. [CrossRef]
20. Rybakov, F.N.; Kiselev, N.S.; Borisov, A.B.; Döring, L.; Melcher, C.; Blügel, S. Magnetic hopfions in solids. *APL Mater.* **2022**, *10*, 111113. [CrossRef]
21. Luk'yanchuk, I.; Tikhonov, Y.; Razumnaya, A.; Vinokur, V.M. Hopfions emerge in ferroelectrics. *Nat. Commun.* **2020**, *11*, 2433. [CrossRef]
22. Thompson, A.; Wickes, A.; Swearngin, J.; Bouwmeester, D. Classification of electromagnetic and gravitational hopfions by algebraic type. *J. Phys. A Math. Theor.* **2015**, *48*, 205202. [CrossRef]
23. Dzyaloshinskii, I.E.; Ivanov, B.A. Localized topological solitons in a ferromagnet. *JETP Lett.* **1979**, *29*, 540–542.
24. Sutcliffe, P. Skyrmion knots in frustrated magnets. *Phys. Rev. Lett.* **2017**, *118*, 247203. [CrossRef] [PubMed]
25. Sutcliffe, P. Hopfions in chiral magnets. *J. Phys. A Math. Theor.* **2018**, *51*, 375401. [CrossRef]
26. Liu, Y.; Hou, W.; Han, X.; Zang, J. Three-dimensional dynamics of a magnetic hopfion driven by spin transfer torque. *Phys. Rev. Lett.* **2020**, *124*, 127204. [CrossRef]
27. Wang, X.S.; Qaiumzadeh, A.; Brataas, A. Current-driven dynamics of magnetic hopfions. *Phys. Rev. Lett.* **2019**, *123*, 147203. [CrossRef]
28. Liu, Y.; Lake, R.K.; Zang, J. Binding a hopfion in a chiral magnet nanodisk. *Phys. Rev. B* **2018**, *98*, 174437. [CrossRef]
29. Tai, J.-S.B.; Smalyukh, I.I. Static Hopf solitons and knotted emergent fields in solid-state non-centrosymmetric magnetic nanostructures. *Phys. Rev. Lett.* **2018**, *121*, 187201. [CrossRef]
30. Göbel, B.; Akosa, C.A.; Tatara, G.; Mertig, I. Topological Hall signatures of magnetic hopfions. *Phys. Rev. Res.* **2020**, *2*, 013315. [CrossRef]
31. Kent, N.; Reynolds, N.; Raftrey, D.; Campbell, I.T.G.; Virasawmy, S.; Dhuey, S.; Chopdekar, R.V.; Hierro-Rodriguez, A.; Sorrentino, A.; Pereiro, E.; et al. Creation and observation of Hopfions in magnetic multilayer systems. *Nat. Commun.* **2021**, *12*, 1562. [CrossRef]
32. Tejo, F.; Hernández Heredero, R.; Chubykalo-Fesenko, O.; Guslienko, K.Y. The Bloch point 3D topological charge induced by the magnetostatic interaction. *Sci. Rep.* **2021**, *11*, 21714. [CrossRef] [PubMed]
33. Guslienko, K.Y. Magnetic vortices and skyrmions. *J. Magn.* **2019**, *24*, 549–567. [CrossRef]
34. Korenman, V.; Murray, J.L.; Prange, R.E. Local-band theory of itinerant ferromagnetism. *Phys. Rev. B* **1977**, *16*, 4032. [CrossRef]
35. Volovik, G.E. Linear momentum in ferromagnets. *J. Phys. C* **1987**, *20*, L83–L87. [CrossRef]
36. Guslienko, K.Y. Gauge and emergent electromagnetic fields for moving magnetic topological solitons. *Europhys. Lett.* **2016**, *113*, 67002. [CrossRef]
37. Faddeev, L.D. *40 Years in Mathematical Physics*; World Scientific Publishing: Singapore, 1995; pp. 374–377.
38. Zheng, F.; Kiselev, N.S.; Rybakov, F.N.; Yang, L.; Shi, W.; Blügel, S.; Dunin-Borkowski, R.E. Hopfion rings in a cubic chiral magnet. *Nature* **2023**, *623*, 718–723. [CrossRef]
39. Yu, X.Z.; Kanazawa, N.; Onose, Y.; Kimoto, K.; Zhang, W.Z.; Ishiwata, S.; Matsui, Y.; Tokura, Y. Near room-temperature formation of a skyrmion crystal in thin-flms of the helimagnet FeGe. *Nat. Mater.* **2011**, *10*, 106–109. [CrossRef]
40. Kundu, A.; Rybakov, Y.P. Closed-vortex-type solitons with Hopf index. *J. Phys. A Math. Gen.* **1982**, *15*, 269–275. [CrossRef]
41. Komineas, S.; Papanicolaou, N. Topology and dynamics in ferromagnetic media. *Phys. D Nonlinear Phenom.* **1996**, *99*, 81–107. [CrossRef]

42. Arfken, G.B.; Weber, H.J. Chapter 1. Vector analysis. In *Mathematical Methods for Physicists*, 6th ed.; Academic Press: San Diego, CA, USA, 2005; pp. 1–95.
43. Khodzhaev, Z.; Turgut, E. Hopfion dynamics in chiral magnets. *J. Phys. Condens. Matter* **2022**, *34*, 225805. [CrossRef]
44. Hobart, R.H. On the instability of a class of unitary field models. *Proc. Phys. Soc.* **1963**, *82*, 201–203. [CrossRef]
45. Derrick, G.H. Comments on nonlinear wave equations as models for elementary particles. *J. Math. Phys.* **1964**, *5*, 1252–1254. [CrossRef]
46. Dzyaloshinskii, I.E. Theory of helicoidal structures in antiferromagnets. *Sov. Phys. JETP* **1964**, *19*, 960–971.
47. Bogdanov, A.N.; Yablonskii, D.A. Thermodynamically stable "vortices" in magnetically ordered crystals. The mixed state of magnets. *Sov. Phys. JETP* **1989**, *68*, 101–103.
48. Skyrme, T.H.R. A non-linear field theory. *Proc. R. Soc. Lond. Ser. A Math. Phys. Sci.* **1961**, *260*, 127–138. [CrossRef]
49. Bogdanov, A. New localized solutions of the nonlinear field equations. *JETP Lett.* **1995**, *62*, 247–251.
50. Guslienko, K.Y. Skyrmion state stability in magnetic nanodots with perpendicular anisotropy. *IEEE Magn. Lett.* **2015**, *6*, 4000104. [CrossRef]
51. Berganza, E.; Jaafar, M.; Fernandez-Roldan, J.A.; Goiriena-Goikoetxea, M.; Pablo-Navarro, J.; García-Arribas, A.; Guslienko, K.; Magén, C.; De Teresa, J.M.; Chubykalo-Fesenko, O.; et al. Half-hedgehog spin textures in sub-100 nm soft magnetic nanodots. *Nanoscale* **2020**, *12*, 18646–18653. [CrossRef]
52. Berganza, E.; Fernandez-Roldan, J.A.; Jaafar, M.; Asenjo, A.; Guslienko, K.; Chubykalo-Fesenko, O. 3D quasi-skyrmions in thick cylindrical and dome-shape soft nanodots. *Sci. Rep.* **2022**, *12*, 3426. [CrossRef]
53. Guslienko, K. 3D Magnetization Textures: Toroidal Magnetic Hopfion Stability in Cylindrical Samples. *Nanomaterials* **2024**, *14*, 125. [CrossRef]
54. Sobucki, K.; Krawczyk, M.; Tartakivska, O.; Graczyk, P. Magnon spectrum of Bloch hopfion beyond ferromagnetic resonance. *APL Mater.* **2022**, *10*, 091103. [CrossRef]
55. Bo, L.; Ji, L.; Hu, C.; Zhao, R.; Li, Y.; Zhang, J.; Zhang, X. Spin excitation spectrum of a magnetic hopfion. *Appl. Phys. Lett.* **2021**, *119*, 212408. [CrossRef]
56. Raftrey, D.; Fischer, P. Field-Driven Dynamics of Magnetic Hopfions. *Phys. Rev. Lett.* **2021**, *127*, 257201. [CrossRef] [PubMed]
57. Liu, Y.; Watanabe, H.; Nagaosa, N. Emergent magnetomultipoles and nonlinear responses of a magnetic hopfion. *Phys. Rev. Lett.* **2022**, *129*, 267201. [CrossRef]
58. Guslienko, K.Y. Magnetic vortex core string gyrotropic oscillations in thick cylindrical dots. *Magnetism* **2022**, *2*, 239–250. [CrossRef]
59. Guslienko, K.Y.; Gareeva, Z.V. Magnetic skyrmion low frequency dynamics in thin circular dots. *J. Magn. Magn. Mater.* **2017**, *442*, 176–182. [CrossRef]
60. Pershoguba, S.S.; Andreoli, D.; Zang, J. Electronic scattering off a magnetic hopfion. *Phys. Rev. B* **2021**, *104*, 075102. [CrossRef]
61. Popadiuk, D.; Tartakovskaya, E.; Krawczyk, M.; Guslienko, K. Emergent magnetic field and non-zero gyrovector of the toroidal magnetic Hopfion. *Phys. Status Solidi (RRL)–Rapid Res. Lett.* **2023**, *17*, 2300131. [CrossRef]
62. Yu, X.; Liu, Y.; Iakoubovskii, K.V.; Nakajima, K.; Kanazawa, N.; Nagaosa, N.; Tokura, Y. Realization and Current-Driven Dynamics of Fractional Hopfions and Their Ensembles in a Helimagnet FeGe. *Adv. Mater.* **2023**, *35*, 2210646. [CrossRef]
63. Knapman, R.; Tausendpfund, T.; Díaz, S.A.; Everschor-Sitte, K. Spacetime magnetic hopfions from internal excitations and braiding of skyrmions. *Commun. Phys.* **2024**, *7*, 151. [CrossRef]
64. Kobayashi, M.; Nitta, M. Winding Hopfions on $\mathbf{R}^2 \times S^1$. *Nucl. Phys. B* **2016**, *876*, 605–618. [CrossRef]
65. Saji, C.; Troncoso, R.E.; Carvalho-Santos, V.L.; Altbir, D.; Nunez, A.S. Hopfion-Driven Magnonic Hall Effect and Magnonic Focusing. *Phys. Rev. Lett.* **2023**, *131*, 166702. [CrossRef]

magnetism

Article

Can We Still Find an Ideal Memristor?

Frank Zhigang Wang

School of Computing, University of Kent, Canterbury CT2 7NF, UK; f.z.wang@kent.ac.uk

Abstract: In 1971, Chua defined an ideal memristor that links magnetic flux φ and electric charge q. In a magnetic lump with a current-carrying conductor, we found that the direct interaction between physical magnetic flux φ and physical electric charge q is memristive by nature in terms of a time-invariant φ-q curve being nonlinear, continuously differentiable and strictly monotonically increasing. Although we succeeded in demonstrating that the "ideal/real/perfect/... memristor" needs magnetism, the structure still suffers from two serious limitations: 1. a parasitic "inductor" effect and 2. bistability and dynamic sweep of a continuous resistance range. Then, we discussed how to overcome these two limitations to make a fully functioning ideal memristor with multiple or an infinite number of stable states and no parasitic inductance. We then gave a number of innovations to the current memristor structure, such as an "open" structure, nanoscale size, magnetic materials with cubic anisotropy (or even isotropy) and sequential switching of the magnetic domains. Contrary to the conjecture that "an ideal memristor may not exist or may be a purely mathematical concept", we remain optimistic that an ideal memristor will be discovered in nature or will be made in the laboratory. Our finding of the memristive flux–charge interaction may advance the development and application of the memristor technology.

Keywords: magnetism; magnetic lump; ideal memristor; neuromorphic computing; brain-inspired computer

1. Introduction

Chua defined the ideal memristor in 1971, which directly interacts with physical magnetic flux φ and physical electric charge q [1], analogous to the resistor, which directly interacts with physical voltage and physical current; the capacitor, which directly interacts with physical voltage and physical charge; and the inductor, which directly interacts with physical current and physical flux.

In 2008, HP made a memristor in titanium dioxide, in which the dopants tend to drift in the direction of the current [2]. Such mobility makes a memristor change resistance. However, from a rigorous theoretical physics perspective, the HP memristor is too incomplete (no magnetic flux), too complex (a "sandwich" structure) and too specialised (even a chemical reaction in the memory-holding oxygen vacancies) to be fundamental.

The major concern is that the HP memristor lacks a magnetic flux term in the original memristor definition. Williams from HP thought the actual definition of memristance would be more general. Williams argued, "Linking electric charge and magnetic flux is one way to satisfy the definition, but it's not the only one. In fact, it turns out you can bypass magnetic interaction altogether." [3].

Can we truly bypass magnetic interaction to define a memristor?

Unfortunately, thus far, it has been somewhat popular to define "φ" and "q" in a purely mathematical way without giving "φ" and "q" any physical interpretations: "φ" was defined as the time integral of voltage "v" and "q" was defined as the time integral of current "i". Most people simply "abused" Chua's suggested fingerprint [4]. "If it (the v-i hysteresis loop) is pinched, it is a memristor" by ignoring its prefixing (from an experimental perspective) and its context ("This definition greatly broadens the scope of memristive devices... into three classes: Ideal Memristors, Generic Memristors, and Extended Memristors.") [4].

At least three examples can be given to demonstrate that it is highly risky to define a memristor along the *v-i* plane.

The first example is indefinite integration itself. Starting from v, we have $\int v(t)dt = \varphi(t) + C$, where C is an arbitrary constant [meaning that any value for C makes $\varphi(t) + C$ a valid antiderivative of $v(t)$]. This constant expresses an ambiguity inherent in the construction of integration; there is an indefinite number of antiderivatives of $v(t)$, and therefore, a given *v-i* curve (as postulated by Chua [1], i is a pinched hysteresis loop (looking like a diagonal "∞")) cannot uniquely determine the *φ-q* curve, as shown in the inset of Figure 1.

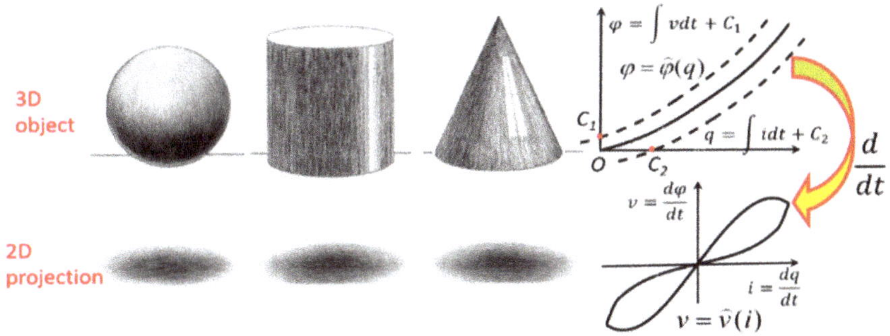

Figure 1. Conformal differential transformation is similar to 3D object projection in terms of projecting something from a "high-dimensionality" space into a "low-dimensionality" space.

In other words, the *φ-q* plane is a space with "high-dimensionality" (if the charge–flux relationship of a memristor is a high-order polynomial as imagined by Chua [5]), whereas the *v-i* plane is its "dimensionality-reduced" space. Conformal differential transformation [4,5] projecting the *φ-q* curve onto the *v-i* plane just plays the role of dimensionality reduction (e.g., $\frac{dx^n}{dx} = nx^{n-1}$). Imagine that we are shining a light from above a 3D object and looking at the shadow it casts on a 2D screen. The transformation from 3D to 2D is unique, but the opposite is not; we cannot work out whether the original 3D object is a ball, cylinder or cone when the 2D projection viewed is a circle. A schematic (Figure 1) best illustrates this. We cannot simply restore the lost information in the projection in a single "anti-projection".

The second example is that, as shown in Figure 2, no hysteresis can be seen in the *v-i* curve if the *φ-q* curve is odd-symmetric, although it complies with all the three criteria [1] for the ideal memristor's *φ-q* curve: 1. nonlinear, 2. continuously differentiable and 3. monotonically increasing. The *φ-q* curve projects to the corresponding *v-i* curve via the so-called conformal differential transformation (for a periodic input current): obtain the angle of incline α of the tangent line at an operating point in the $\varphi = \hat{\varphi}(q)$ curve and draw a straight line through the origin on the *v-i* plane whose angle of incline is also α.

The third example is that, as shown in Figure 3, a pinched *v-i* curve that is asymmetric results from even a double-valued *φ-q* curve, which indicates that the corresponding memristor is not ideal at all. It is concluded that an ideal memristor that is originally defined on the constitutive *φ-q* plane should not be characterized on the *v-i* plane.

What we have observed from the above three examples is that "even if it's pinched it may not be an ideal memristor". In our opinion, it is bad to define the ideal memristor on the *v-i* plane, it is worse to have no physical magnetic flux, and it is even worse to pretend to have a magnetic flux that is virtual and calculated from other physical attributes. Strictly speaking, it is incorrect to bypass or replace the direct, physical charge–flux interaction with a "virtual" interaction between the charge (that is normally physical, e.g., oxygen vacancies as ionic current in the HP memristor [2]) and the integral over the voltage (having no choice due to the lack of magnetic flux).

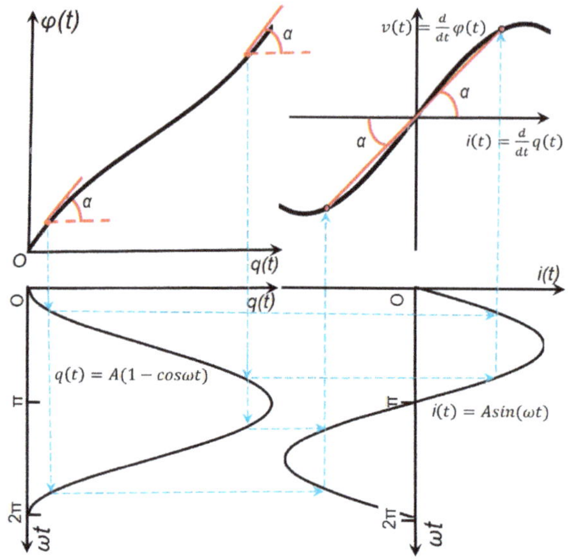

Figure 2. No hysteresis can be seen in the v-i curve for an ideal memristor with an odd-symmetric φ-q curve.

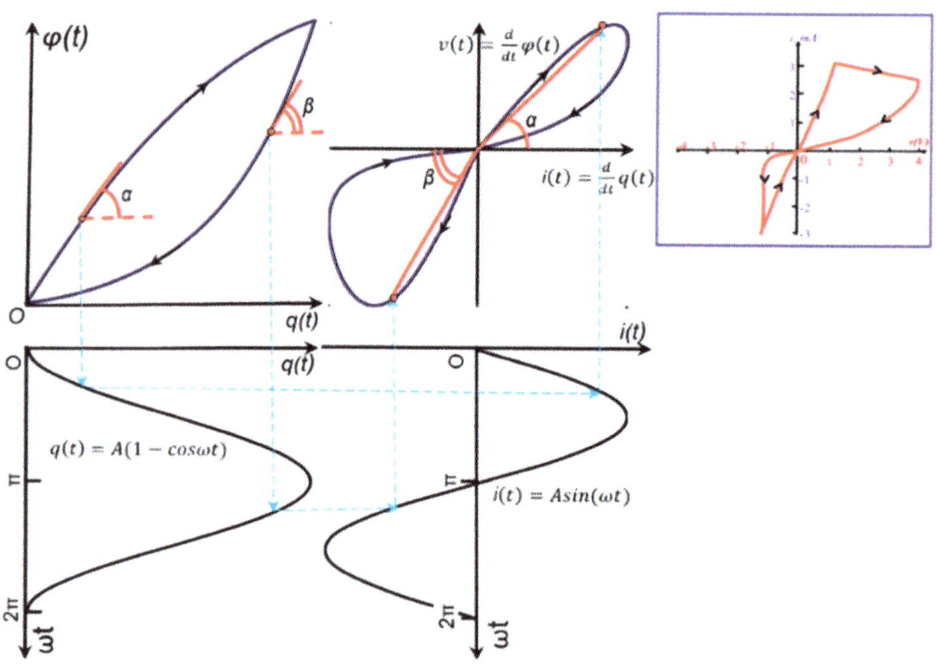

Figure 3. A nonideal memristor even has a pinched v-i hysteresis loop. This memristor has two φ-q characteristic branches, each of which is chosen depending on the polarity of the input current. Some practical devices exhibit such q–φ curves [6].

Other sceptics have also expressed a similar concern regarding the lack of a charge–flux interaction in the HP memristor. As mentioned in the same issue of IEEE Spectrum, as early as in 2009, sceptics have argued that the HP memristor is not a fourth fundamental circuit element but an example of bad science [7]. In 2015, Vongehr even declared, "The Missing Memristor Has Not Been Found" (the title of their Nature Scientific Reports paper) in the sense that an ideal memristor device should be grounded in fundamental symmetries of basic physics, here electromagnetism and that the "ideal/real/perfect/... memristor" needs magnetism [8]. It is worth mentioning that Vongehr's work [8] is only one facet of the opposition to the ideality of a new passive fundamental electrical component. In 2017, Sundqvist, et al. showed that the claim that the present memristor functions describe a passive device violates the second law of thermodynamics in many cases [9,10]. In 2018, Abraham proposed a case to reject the memristor as a fundamental circuit element [11]. In 2022, Kim, et al. refuted the claim that all resistance-switching memories are memristors and doubted the existence of ideal memristors as actual physical devices that can be fabricated experimentally [12].

2. Magnetic Lump with Flux–Charge Interaction

In order to design an artificial device with a direct interaction between physical magnetic flux and physical electric charge and then prove that the interaction is memristive, the first principles originating from Aristotelians more than 2300 years ago may help. That is to say, we should start directly at the level of the established memristor concept based on physical charge and physical flux without making any assumption such as an empirical model (e.g., a pinched hysteresis voltage–current loop) and parameter fitting (e.g., voltage as a derivative of flux and current as a derivative of the charge).

After numerous failures, the structure in Figure 4 was eventually designed to introduce a direct charge–flux interaction. Through the conductor that carries a current and the magnetic lump that hosts magnetization, the amount of the charge can be controlled by the time interval of the current flow. The flux can be adjusted by the magnetization rotation. In order to pick up a signal from the charge–flux coupling, this conductor can simultaneously play two roles; on the one hand, it carries a current to switch the magnetization in the lump, and on the other hand, it can also sense the voltage possibly induced by the switched flux.

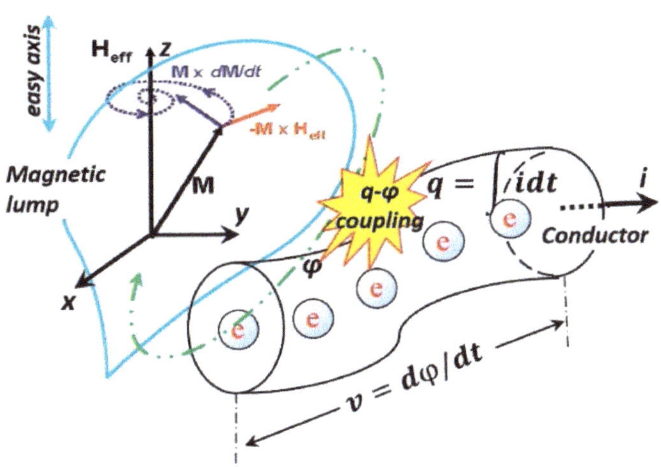

Figure 4. The flux–charge interaction in a structure with a magnetic lump and a current-carrying conductor. The Oersted field generated by the current i rotates or switches the magnetization M inside the magnetic lump, and consequently, the switched magnetic flux φ induces a voltage v across the conductor, resulting in a changed (equivalent) memristance. The LLG model of flux reversal is also shown. If the magnetic field H_{eff} is applied in direction Z, the saturation magnetization vector ***M(t)*** follows a precession trajectory from its initial position ($m_0 \approx -1$) until ($m \approx 1$), i.e., the magnetization ***M(t)*** reverses itself and is eventually aligned with the magnetic field H_{eff}.

3. LLG Model of Magnetic Flux Reversal

The resulting equivalent resistance, the ratio between the sensed voltage and the driving current, of the above structure is possibly memristive. It would be highly unlikely that it would have a linear charge–flux interaction due to the existence of magnetic material with rich hysteresis. That is, it is the magnetic lump material that may provide a source of nonlinearity necessary for such a structure.

In order to describe the charge–flux interaction in a mathematically and physically rigorous way, the Landau–Lifshitz–Gilbert equation [13,14] was used. Assuming $m(t) = \frac{M_Z(t)}{M_S}$, in which M_Z is the component of the saturation magnetization M_S in the Z axis, and a magnetic field H is applied along Z. The model is expressed as below:

$$m(t) = tanh\left[\frac{q(t)}{S_W} + C\right], \tag{1}$$

in which S_W is a switching coefficient, and C is a constant of integration such that $C = tanh^{-1}m_0$ (m_0 is the initial value of m) if $q(t = 0) = 0$ (no accumulation of charge at any point).

By Faraday's law, the induced voltage $v(t)$ is

$$\mu_0 S\frac{dM_z}{dt} = S\frac{dB_z}{dt} = \frac{d\varphi_z}{dt} = -v(t), \tag{2}$$

where μ_0 is the permeability of free space and S is the cross-sectional area.

From Equation (2), we obtain

$$\varphi = \mu_0 SM + C' = \mu_0 SM_S m + C', \tag{3}$$

where C' is another constant of integration.

Assuming $\varphi(t = 0) = 0$, we have $C' = -\mu_0 SM_S m_0$; so,

$$\varphi = \mu_0 SM_s\left[\tanh\left(\frac{q}{S_W} + tanh^{-1}m_0\right) - m_0\right] \triangleq \hat{\varphi}(q). \tag{4}$$

Equation (4) complies with the three new criteria [6] for the ideal memristor: 1. nonlinear; 2. continuously differentiable; and 3. strictly monotonically increasing. Figure 5 shows a typical φ-q curve with $m_0 = -0.964$ (such a value reflects the intrinsic fluctuation; otherwise, $\boldsymbol{M}$ will stick to the stable equilibriums $m_0 = \pm 1$).

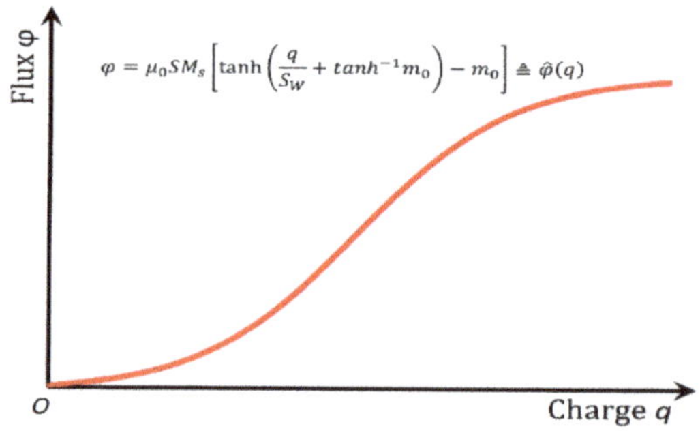

Figure 5. Intuitively, the S-shaped φ-q curve (Equation (4)) of this structure vividly depicts the self-limiting charge–flux interaction in a circuit element. It complies with the three new criteria for the ideal memristor [1,6]: a. nonlinear; b. continuously differentiable; c. strictly monotonically increasing.

Figure 5 agrees with those experimentally observed φ-q curves in the magnetic cores [15–17]. Therefore, Equation (4) is used as a constitutive curve in this work. The constitutive curve in Figure 4 is dramatically different from that of Chua's "fictitious" memristor with a charge–flux relationship $\varphi = \frac{1}{3}q^3 + q$ in his tutorial [5]. We have $\lim_{q\to\pm\infty} \tanh(q) = \pm 1$ for a hyperbolic function whose output range is normalized from -1 to 1 (no matter how big the input is), whereas we have $\lim_{q\to\pm\infty}\left(\frac{1}{3}q^3 + q\right) = \pm\infty$ for a polynomial function (that diverges to infinity). In other words, such a structure's operation range is finite (where $M(q) = \frac{d\varphi}{dq} \neq 0$), whereas the "fictitious" memristor's range is infinite. Such a self-limiting hyperbolic tangent function is more natural than other functions since the S curve approaches 1 as x is +∞ and approaches zero as x is $-\infty$. In biology and ecology, a self-limiting colony of organisms limits its own growth by its actions (releasing waste that is toxic to the colony once it exceeds a certain population) [18]. In this instance, there is a clear physical explanation for the saturation of this structure: it is because the magnetization vector is as aligned as the magnetic field allows it to be. The change in the magnetization alignment is negligible on increasing the field above this.

If a step-function excitement current is applied and its rise time is short enough in approaching constant *I*, by Faraday's law, the induced voltage is

$$v(t) = \frac{d\varphi}{dt} = \frac{\mu_0 S M_S I}{S_W} sech^2\left(\frac{I}{S_W}t + tanh^{-1}m_0\right). \quad (5)$$

Equation (5) is depicted in Figure 6.

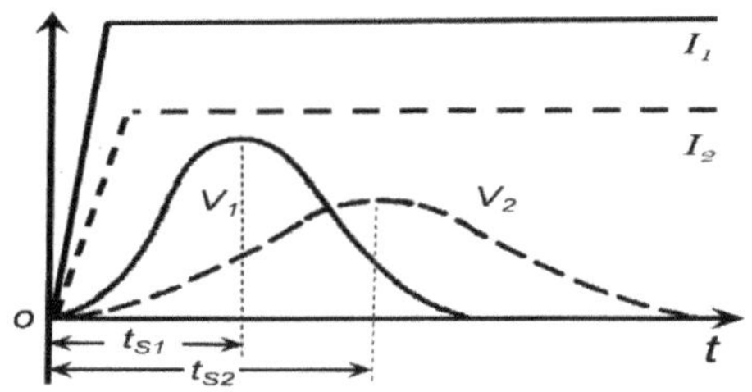

Figure 6. Induced voltage vs. a step-function input current. The higher the amplitude of the current *I*, the shorter the switching time t_S.

From Equation (4), the memristance *M(q)* of our memristor is

$$M(q) = \frac{d\varphi}{dq} = \frac{\mu_0 S M_s}{S_W} sech^2\left(\frac{q}{S_W} + tanh^{-1}m_0\right) \geq 0. \quad (6)$$

4. Parasitic Inductance and Stepwise Memristance

Inevitably, parasitic inductance of this structure in Figure 4 coexists, especially if it is a core with a high coupling efficiency (Figure 7), which can be described by

$$L = 0.4\pi\mu N^2 \frac{A}{l} \times 10^{-5}(mH), \quad (7)$$

where L is the inductance (mH) of the magnetic core, $A = \frac{D-d}{2} \cdot h\ (cm^2)$ is the cross-sectional area of the core, $l = \pi \cdot \frac{D+d}{2}\ (cm)$ is the average length of the core, μ is the permeability of the core, N is the number of the turns in the coil ($N = 1$ in Figure 4), $D\ (cm)$ is the outer diameter of the core, $d\ (cm)$ is the inner diameter of the core, and $h\ (cm)$ is the height of the core [19].

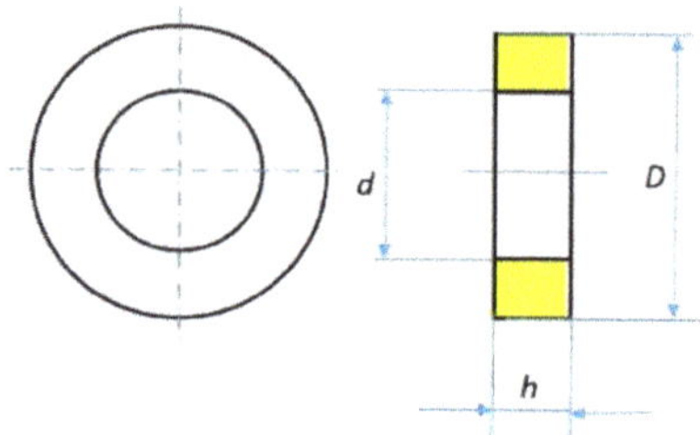

Figure 7. Geometry of a magnetic core for the inductance calculation.

If a fixed aspect ratio $D = 2d = 2h$ is taken, Equation (5) can be simplified to

$$L = 0.4\pi\mu N^2\frac{h}{3} \times 10^{-5}(mH), \quad (8)$$

which implies that the inductance of a magnetic core scales with its physical size. This is truly encouraging in the sense that, in principle, an ideal memristor at the nanoscale is expected to have negligible parasitic inductance.

Notably, as shown in Equation (6), the parasitic inductance in this device is not a function of the charge (otherwise, it will become a mem-inductor [20,21]), whereas its resistance (memristance) is a function of the charge.

An antiferromagnet/ferromagnet heterostructure [22] was found to exhibit a stepwise memristance (due to the sequential switching of the domains in the ferromagnetic layer), in which one can freeze the resistance statically at any intermediate time point. It exhibits a continuous "state-dependent Ohm's law" and provides a solution of Chua's Enigma: All non-volatile memristors have continuum memories [23]. Pershin, Di Ventra and Kim designed a test to check experimentally if a resistor with memory is indeed a memristor, in terms of its resistance depending only on the charge that flows through it or on the history of the voltage across it [24–26].

5. Conclusions and Arguments

Our innovative work represents a step forward in terms of verifying the memristive charge–flux interaction, but we have not reached the final The structure has two serious limitations:

1. The aforementioned memristive fingerprint hides behind a superficial inductor effect due to its inductor-like structure. It was necessary to apply a constant input current (such as a step-function or a sequence of square-wave pulses) to depress the inductor effect ($\because v = L\frac{di}{dt} = 0$). Despite the existence of parasitic inductance, the structure displays memristivity; similarly, a real-world resistor is still thought to be a resistor despite the existence of an (inevitable) parasitic inductance and/or capacitance. Most importantly, the structure exhibits that its charge–flux interaction is memristive by nature.
2. The structure is bistable and dynamically sweeps a continuous range of resistances. This "dynamical continuity" results from the uniaxial magnetic anisotropy of the prototype, which contains magnetic material with only one easy axis. A fully functioning ideal memristor should have multiple or an infinite number of stable states so its static memristance can be "frozen" at any intermediate point in time.

Limitation 2 ("dynamic continuity") may be overcome by using magnetic materials with cubic anisotropy (three or four easy axes as shown in Figure 8 [27]) or even a magnetically isotropic material (no preferential direction for its magnetisation) [28]. In addition, an antiferromagnet/ferromagnet heterostructure exhibits stepwise memristance due to sequential switching of the domains in its ferromagnetic layer [22]. However, this bilayer is nonideal since its resistance is a function of several internal state variables, including temperature, and it is not a function of only charge [22].

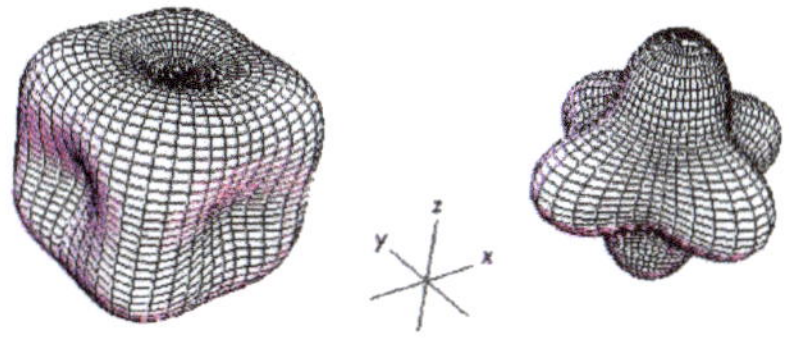

Figure 8. Cubic anisotropy energy surfaces. The magnetic moment in cobalt has a dependence of energy level towards one particular direction (the easy axis); then, it has uniaxial anisotropy. The biases in iron (**left**) and nickel (**right**) are toward many particular directions; then, they have multiple easy axes and possess cubic anisotropy [27].

A fully functioning charge–flux-interaction-based ideal memristor with multiple or an infinite number of stable states and no parasitic inductance is still highly in demand in terms of filling the gap of 50 years [1,7,8] and advancing the development and application of memristor technology. The existence of such a fundamental circuit element may appeal to many researchers in the memristor field within the context of the theoretical circuit innovations that depend on charge–flux linkage [28–32]. We are still optimistic that researchers will discover an ideal memristor in nature or make one in the laboratory, although some researchers feel that an ideal memristor may not exist or may be a purely mathematical concept [25].

Funding: This research was partially funded by an EC grant "Re-discover a periodic table of elementary circuit elements", PIIFGA2012332059, Marie Curie Fellow: Leon Chua (UC Berkeley), Scientist-in-charge: Frank Wang (University of Kent).

Data Availability Statement: All data generated and analyzed during this study are included in this published article.

Conflicts of Interest: The authors declare no conflicts of interest. The funders had no role in the design of the study; in the collection, analyses, or interpretation of data; in the writing of the manuscript; or in the decision to publish the results.

References

1. Chua, L. Memristor—The Missing Circuit Element. *IEEE Trans. Circuit Theory* **1971**, *18*, 507–519. [CrossRef]
2. Strukov, D.; Snider, D.; Stewart, S.; Williams, S. The missing memristor found. *Nature* **2008**, *453*, 80–83. [CrossRef]
3. Williams, R.S. How we found the missing memristor. *IEEE Spectr.* **2008**, *45*, 29–35. [CrossRef]
4. Chua, L. If it's pinched it's a memristor. *Semicond. Sci. Tech.* **2014**, *29*, 104001. [CrossRef]
5. Chua, L. Resistance switching memories are memristors. *Appl. Phys. A* **2011**, *102*, 765–783. [CrossRef]
6. Georgiou, P.S.; Barahona, M.; Yaliraki, S.N.; Drakakis, E.M. On memristor ideality and reciprocity. *Microelectron. J.* **2014**, *45*, 1363–1371. [CrossRef]
7. Adee, S. Resistance to Memristance. *IEEE Spectr.* **2008**, *45*, 34. [CrossRef]
8. Vongehr, S.; Meng, X. The missing memristor has not been found. *Nat. Sci. Rep.* **2015**, *5*, srep11657. [CrossRef]
9. Sundqvist, K.M.; Ferry, D.K.; Kish, L.B. Memristor Equations: Incomplete Physics and Undefined Passivity/Activity. *Fluct. Noise Lett.* **2017**, *16*, 1771001. [CrossRef]
10. Sundqvist, K.M.; Ferry, D.K.; Kish, L.B. Second Law based definition of passivity/activity of devices. *Phys. Lett. A* **2017**, *381*, 3364–3368. [CrossRef]
11. Abraham, I. The case for rejecting the memristor as a fundamental circuit element. *Sci. Rep.* **2018**, *8*, 10972. [CrossRef]
12. Kim, J.; Pershin, Y.; Yin, M.; Datta, T.; Di Ventra, M. A Definitive Demonstration that Resistance-Switching Memories Are Not Memristors. Available online: http://nvmw.ucsd.edu/nvmw2022-program/nvmw2022-data/nvmw2022-paper1-final_version_your_extended_abstract.pdf (accessed on 27 June 2024).
13. Landau, L.D.; Lifshitz, E.M. Theory of the dispersion of magnetic permeability in ferromagnetic bodies. *Physik. Z. Sowjetunion* **1935**, *8*, 153–169.
14. Gilbert, T.L. A Lagrangian formulation of the gyromagnetic equation of the magnetic field. *Phys. Rev.* **1955**, *100*, 1243.
15. Menyuk, N.; Goodenough, J. Magnetic materials for digital computer components. I. A theory of flux reversal in polycrystalline ferromagnetics. *J. Appl. Phys.* **1955**, *26*, 8–18. [CrossRef]
16. Gyorgy, E.M. Rotational model of flux reversal in square loop ferritcs. *J. Appl. Phys.* **1957**, *28*, 1011–1015. [CrossRef]
17. Cushman, N. Characterization of Magnetic Switch Cores. *IRE Trans. Compon. Parts* **1961**, *8*, 45–50. [CrossRef]

18. Available online: https://en.wikipedia.org/wiki/Self-limiting_(biology) (accessed on 27 June 2024).
19. Approximate Inductance of a Toroid. Available online: http://hyperphysics.phy-astr.gsu.edu/hbase/magnetic/indtor.html (accessed on 25 June 2024).
20. Di Ventra, M.; Pershin, Y.V.; Chua, L.O. Memristive circuits simulate memcapacitors and meminductors. *Proc. IEEE* **2009**, *97*, 1717. [CrossRef]
21. Wang, F.Z. Beyond Memristors: Neuromorphic Computing Using Meminductors. *Micromachines* **2023**, *14*, 486. [CrossRef] [PubMed]
22. Kurenkov, A.; DuttaGupta, S.; Zhang, C.; Fukami, S.; Horio, Y.; Ohno, H. Artificial Neuron and Synapse Realized in an Antiferromagnet/Ferromagnet Heterostructure Using Dynamics of Spin–Orbit Torque Switching. *Adv. Mater.* **2019**, *31*, e1900636. [CrossRef]
23. Chua, L. Five non-volatile memristor enigmas solved. *Appl. Phys. A* **2018**, *124*, 563. [CrossRef]
24. Pershin, Y.V.; Kim, J.; Datta, T.; Di Ventra, M. An experimental demonstration of the memristor test. *arXiv* **2021**, arXiv:2102.11963. [CrossRef]
25. Kim, J.; Pershin, Y.V.; Yin, M.; Datta, T.; Di Ventra, M. An experimental proof that resistance-switching memory cells are not memristors. *Adv. Electron. Mater.* **2020**, *6*, 2000010. [CrossRef]
26. Pershin, Y.V.; Di Ventra, M. A simple test for ideal memristors. *J. Phys. D Appl. Phys.* **2018**, *52*, 01LT01. [CrossRef]
27. Aharoni, A. *Introduction to the Theory of Ferromagnetizm*, 2nd ed.; Clarendon Press: London, UK, 1996; pp. 330–336.
28. Biolek, D.; Vavra, J.; Biolek, Z.; Kolka, Z.; Biolkova, V.; Dobes, J. Chua's Table as a Tool for Constructing Dual Networks. In Proceedings of the 2019 IEEE Asia Pacific Conference on Circuits and Systems (APCCAS), Bangkok, Thailand, 11–14 November 2019.
29. Biolek, D.; Biolek, Z.; Biolkova, V.; Kolka, Z. X-Controlled Memristive Devices for Automatic Gain Control in RC Oscillators. In Proceedings of the 2020 New Trends in Signal Processing, Demanovska Dolina, Slovakia, 14–16 October 2020.
30. Eshraghian, J.K.; Cho, K.-R.; Iu, H.H.C.; Fernando, T.; Iannella, N.; Kang, S.-M.; Eshraghian, K. Maximization of crossbar array memory using fundamental memristor theory. *IEEE Trans. Circuits Syst. II Express Briefs* **2017**, *64*, 1402–1406. [CrossRef]
31. Eshraghian, J.K.; Iu, H.H.; Fernando, T.; Yu, D.; Li, Z. Modelling and characterization of dynamic behavior of coupled memristor circuits. In Proceedings of the 2016 IEEE International Symposium on Circuits and Systems (ISCAS), Montreal, QC, Canada, 22–25 May 2016.
32. Chua, L. Everything You wish to know about memristors but are afraid to ask. In *Handbook of Memristor Networks*; Springer: Berlin/Heidelberg, Germany, 2019.

magnetism

MDPI

Article

Loss Mitigation in Self-Biased Microstrip Circulators

Lingqi Kong [1], Alexander Schuchinsky [1,*], Sumin Joseph [2], Taylan Eker [3] and Yi Huang [1,*]

1 Department of Electrical Engineering and Electronics, The University of Liverpool, Liverpool L7 3QE, UK
2 Department of Electronic and Electrical Engineering, The University of Sheffield, Sheffield S1 3JD, UK
3 Huawei Technologies Sweden AB, 4 Skalholtsgatan 9-11, SE-164 40 Kista, Sweden
* Correspondence: aschuch@liverpool.ac.uk (A.S.); yi.huang@liverpool.ac.uk (Y.H.)

Abstract: Integration of the ferrite devices in the RF front-end and active antennas is hindered by the need for external magnets, biasing soft microwave ferrites. The hexaferrite-based self-biased nonreciprocal devices can operate without external magnets at mm-wave frequencies but the currently available hexaferrite materials inflict high RF losses at lower frequencies, particularly in the wireless communication bands. In this paper, the parameters of La-Co-substituted hexaferrite compounds are used for the self-biased circulators in the low GHz frequency bands, and a means of the dissipation loss reduction are discussed.

Keywords: hexaferrite; magnetic loss; magnet-free nonreciprocal device; self-biased circulator

1. Introduction

Nonreciprocal passive devices are indispensable elements of the high-performance transceivers in radio frequency (RF) front ends of communication systems and radars [1–3]. Circulators and isolators protect the very sensitive front end of the receivers, operating in a duplex mode with transmitters connected to the same antenna.

Conventional passive nonreciprocal microwave components employ magnetically soft ferrites with a narrow linewidth of ferrimagnetic resonance. They are biased by external permanent magnets [4,5]. However, magnets make the ferrite devices bulky and difficult to package with other RF components and antennas in modern highly integrated electronic systems. Magnets also cause high RF losses and interfere with surrounding densely packed electronic circuits. Therefore, low-loss magnet-free circulators and isolators compatible with modern fabrication technologies are a long-standing goal [5].

Hexaferrites with strong magneto-crystalline anisotropy have been the primary candidates for realizing magnet-free nonreciprocal devices; see [6] and references therein. However, the high RF losses of hexaferrite materials remain the main obstacle. Recent advances in the synthesis of magnetically hard La-Co-substituted hexaferrite compounds have reinvigorated efforts in the development of miniature self-biased low-loss circulators and isolators [7–11]. The magnet-free microstrip circulators were reported to achieve a fractional bandwidth (FBW) of 3% and an insertion loss (IL) of 1.52 dB at a centre frequency of 13.65 GHz [8] and 1 dB losses at 30 GHz [10]. A lower IL of 0.87 dB was achieved in the waveguide junction circulator with a hexaferrite post [11] at the centre frequency of 41 GHz. Additionally, its IL dramatically reduced to 0.21 dB when an external magnetic bias was applied to the hexaferrite post of this circulator. Such a major improvement of the IL implies that the high losses are not endemic to the hexaferrite-based devices and they can be significantly decreased by conditioning the internal magnetic field of the self-biased hexaferrite junction resonator. However, the conventional hexaferrite materials are not suitable for low GHz frequencies, and high performance self-biased circulators are not commercially available for frequencies below 26 GHz [12].

An alternative concept of magnet-free circulators and isolators is based on the inherent nonreciprocity of voltage-biased FET and varactor diodes [13–16]. The published simulation and experimental results have proved the principle, but the ILs of such devices achieved to date still exceed 3 dB and remain notably higher than the IL of the hexaferrite-based devices. It is also necessary to stress that the inherent nonlinearity of the semiconductor-based passive nonreciprocal devices imposes additional constraints on their use in the high-performance RF front ends exposed to the high RF power.

In this paper, the mechanisms of RF loss in self-biased microstrip circulators on hexaferrite substrates are examined and a means of a significant loss reduction are discussed. The properties and performance of a La-Co-substituted hexaferrite composition tailored for Ku band operation are outlined in Section 2. In Section 3, the sources of RF loss in self-biased circulators are discussed and a low-loss circulator with a central frequency of 16 GHz is described. It is demonstrated that the RF losses can be reduced by more than half, to the level below 0.66 dB, and the fractional bandwidth increases to about 5% when the matching transformers are located on the alumina substrate surrounding the hexaferrite junction resonator. The performance of the self-biased microstrip circulators with the operating frequencies of 4 and 6 GHz are evaluated and discussed in Section 4. The main results are summarized in the Conclusion.

2. La-Co Hexaferrites for Circulator Junction Resonators

Hexaferrites with high remanent magnetization, $4\pi M_r$, and low magnetic loss are the key elements of magnet-free passive nonreciprocal RF devices. Unlike soft microwave ferrites, which require permanent magnets for external DC magnetic bias, hexaferrites are self-biased and retain their magnetisation. They contain two sublattices:

- Hard sublattice provides a magnetic bias, like an external permanent magnet.
- Soft sublattice acts similarly to the soft microwave ferrite.

Thus, the soft lattice is biased internally by the DC magnetic field of the hard lattice. As a result, at RF frequencies, hexaferrite can behave like a conventional soft ferrite biased by the DC magnetic field of its hard lattice. The gyrotropic properties of hexaferrite compounds are rendered by the strong coercive magnetic field $H_c > 4\pi M_r$ created by the uniaxial magneto-crystalline anisotropy. A particular hexaferrite composition is dictated by the device operating frequency and the trade-offs between magneto-crystalline anisotropy, the demagnetizing field and ferrimagnetic resonance linewidth ΔH. A range of M-type- and SM-type-substituted hexaferrites has been developed for high RF applications [6].

2.1. Properties of La-Co-Substituted Hexaferrites

The La-Co-substituted hexaferrite compounds of type $Sr_{1-x}La_xFe_{12-x}Co_xO_{19}$ have been recently developed for RF applications. The salient features of these compositions include the possibility of adjusting the magneto-crystalline anisotropy by varying the substitution rate x up to 0.4 with negligible effect on the high value of $4\pi M_r$ and Curie temperature which remains above 400 °C. This allows the hexaferrite parameters to be tailored for the specific operational frequency bands.

Polycrystalline hexaferrites can be produced using the ceramic fabrication process [17]. Their raw ingredients, $SrCo_3$, Fe_2O_3, La_2O_3, and Co_3O_4, are milled and mixed together first. Then the mixture is calcinated at a temperature between 1000 °C and 1100 °C. The produced powder is finely milled again and sintered at temperatures of 1200–1300 °C. A crystallographic orientation of the specimens is enforced by the DC magnetic field, H_{ex}, applied when the calcinated powder is pressed. A high rate of crystallite alignment is facilitated by the use of very fine powder with submicron particles and a strong magnetic bias H_{ex} exceeding 10 kOe.

Disk-shaped hexaferrite specimens of a diameter of 1.74 mm and thickness of 200 μm were characterized by superconducting quantum interference device (SQUID) magnetometer [18]. The measurements were made along the sample's principal magnetic axes: the easy c-axis and the hard axes a and b, defined in Figure 1. The magnetization was measured versus H_{ex} varied up to 70 kOe to ensure that the magnetically hard hexaferrite crystallites are fully aligned and saturated along each axis.

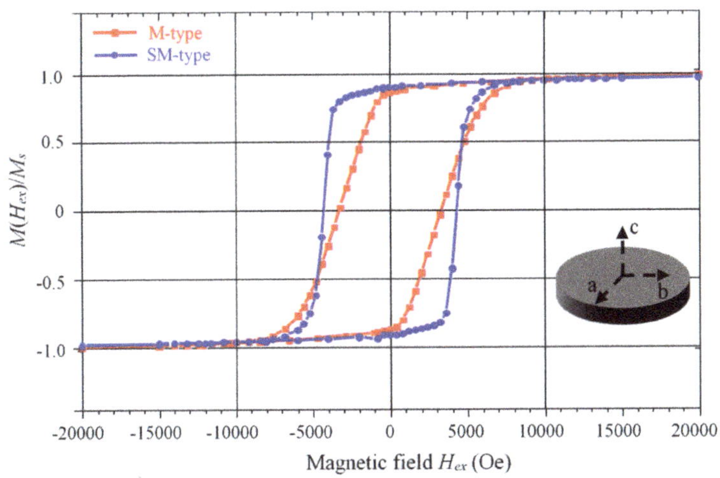

Figure 1. Hysteresis curves of magnetization $4\pi M(H_{ex})$ normalized to the saturation magnetization $4\pi M_s \approx 4500$ Gs for the M-type (red line) and SM-type (blue line) hexaferrites. DC magnetic field H_{ex} is directed along the easy c-axis.

The measured hysteresis curves of the magnetization $M(H_{ex})$ normalized to the saturation magnetization $4\pi M_s$ are shown in Figure 1 for the base compound $SrFe_{12}O_{19}$ and the substituted hexaferrite $Sr_{0.8}La_{0.2}Fe_{11.8}Co_{0.2}O_{19}$ ($x = 0.2$), labelled M and SM, respectively. It is evident in Figure 1 that the La-Co-substituted SM-type hexaferrite magnetized along the c-axis has a squarer hysteresis loop than the M-type hexaferrite. The higher remanent magnetization of the SM-type hexaferrite is the result of its stronger coercive magnetic field, H_c. Its resilience to demagnetization is especially beneficial for the homogeneity of H_c, which serves as an internal DC magnetic bias H_i of the hexaferrite.

An average magnetization of the hexaferrite substrate of the SM-type was measured by the SQUID magnetometer, and the effective linewidth ΔH_{eff} was retrieved from the waveguide measurements similar to [19]. The hexaferrite permittivity was determined separately by using the coaxial line [20]. The obtained physical parameters of the hexaferrite substrate are summarized in Table 1, where $4\pi M_s$ is the saturation magnetization; $4\pi M_r$ is the remanent magnetization; H_c is the coercive magnetic field (internal magnetic bias); ε_f and tan δ_f are relative permittivity and dielectric loss tangent; ΔH is the ferrimagnetic resonance linewidth measured at frequency $f_r = 53.48$ GHz; and Δ_{eff} is the effective linewidth evaluated at frequency 16 GHz. These parameters are used for modelling and analysis of the self-biased junction circulators in Section 3.

Table 1. Parameters of La-Co-substituted SM-type hexaferrite substrate.

$4\pi M_s$, kGs	$4\pi M_r$, kGs	H_c, kOe	ε_f	tan δ_f	ΔH, Oe	ΔH_{eff}, Oe
4.5	4.0	19.1	25	0.002	1000	20

2.2. DC Magnetic Field Profile in Hexaferrite Slab

Distributions of the DC magnetic field H_{DC} on the surfaces of a thin slab of SM-type hexaferrite were mapped by the Hall sensor with an active area of 1.0×0.5 mm^2. A sample with a surface area of 5.0×6.0 mm^2 and thickness of 250 ± 3 μm was magnetized to saturation along the c-axis normal to the layer surface.

A contour plot of the normal component of H_{DC} mapped at the height of 3.5 mm above the surface is shown in Figure 2. The measured H_{DC} distributions are similar on both sides of the layer, but the field magnitudes differ slightly due to minor differences in the probe positioning. The measured H_{DC} is fairly uniform above the specimen central area but gradually decreases towards the edges. Such a pattern of H_{DC} suggests that the internal magnetic bias, H_i, in the hexaferrite layer is reasonably homogeneous.

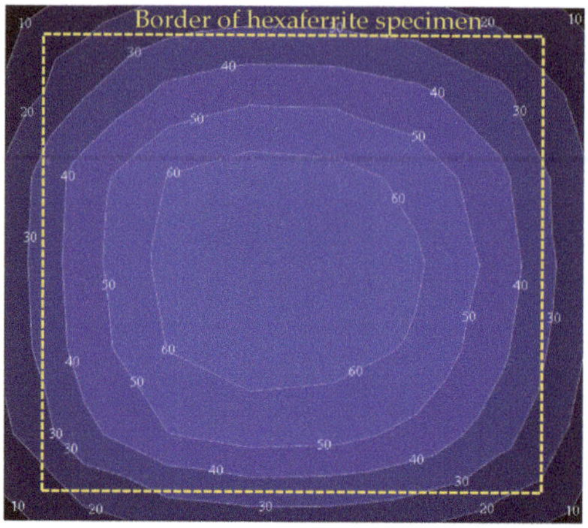

Figure 2. Pattern of the normal component of DC magnetic field H_{DC} measured in Oe at the height of 3.5 mm above the hexaferrite slab with a surface area of 5.0×6.0 mm^2 and thickness of 250 μm.

It is necessary to stress that the homogeneity of H_i and remanent magnetization, $4\pi M_r$, are the essential prerequisites for the low-loss operation of nonreciprocal ferrite devices. Therefore, La-Co-substituted SM-type hexaferrite slab with inherently strong magneto-crystalline anisotropy, high $4\pi M_r$ and strong H_c is particularly apt for these applications. Small spatial variations of the demagnetizing field due to the shape of the hexaferrite specimens can be mitigated similarly to [11] where the inhomogeneous DC magnetic bias was applied externally to compensate for the effects of shape and aspect ratio of the hexaferrite specimen.

3. Self-Biased Microstrip Junction Circulators

The junction resonator on the hexaferrite substrate is the core element of the self-biased circulator enabling its nonreciprocal response. Hexaferrite, magnetized along the z-directed c-axis, is described by the Polder tensor of permeability [21]

$$\boldsymbol{\mu} = \mu_0 \begin{bmatrix} \mu & -\mathrm{j}\kappa & 0 \\ \mathrm{j}\kappa & \mu & 0 \\ 0 & 0 & 1 \end{bmatrix} \tag{1}$$

where

$$\mu = 1 + \frac{\omega_0 \omega_{\mathrm{m}}}{\omega_0^2 - \omega^2} \tag{2}$$

$$\kappa = \frac{\omega \omega_{\mathrm{m}}}{\omega_0^2 - \omega^2} \tag{3}$$

$$\omega_0 = \gamma(H_i + j\Delta H) \tag{4}$$

$$\omega_{\mathrm{m}} = \gamma 4\pi M_r \tag{5}$$

ω is an angular frequency, $4\pi M_r$ is the remanent magnetization, ω_0 is the complex-valued angular frequency of ferrimagnetic resonance, $H_i = H_c$ is the internal DC magnetic bias equal to the average magnitude of the coercive magnetic field, imposed by the magneto-crystalline anisotropy of hexaferrite, ΔH is the ferrimagnetic resonance linewidth, and $\gamma = 2.8$ GHz/kOe is the gyromagnetic ratio.

The geometry of hexaferrite-based Y-junction circulators can be initially approximated with the aid of the models developed in [22–24] for the stripline circulators with externally biased ferrite resonators. Bosma's model [22] defines the approximate relation between circulator operating frequency ω, radius R of the disk-shaped junction resonator, and the hexaferrite relative permittivity ε_f and effective permeability μ_e

$$kR\sqrt{\varepsilon_f\mu_e} \approx 1.84 \tag{6}$$

where $k = \omega/c$ is the free space wavenumber, c is the speed of light in free space,

$$\mu_e = \frac{\mu^2 - \kappa^2}{\mu} = \frac{(\omega_0 + \omega_m)^2 - \omega^2}{\omega_0(\omega_0 + \omega_m) - \omega^2} \tag{7}$$

is an effective permeability and the frequency of the transverse ferrimagnetic resonance

$$\omega_\perp = Re\left(\sqrt{\omega_0(\omega_0 + \omega_m)}\right) \tag{8}$$

Numerical estimates of (2), (3) and (7) show that at the hexaferrite parameters specified in Table 1, $\mu \approx 1.23 + 0.002j$, $\kappa \approx 0.069 + 0.002j$, $\mu_e \approx 1.226 + 0.002j$ are in the middle of the Ku band. It is noteworthy that the magnetic losses of hexaferrite at these frequencies are commensurate to the dielectric losses despite ΔH of the hexaferrite is much larger than that of the soft ferrites. This is the result of the operating frequency offset from the transverse ferrimagnetic resonance (8) at $\omega_\perp/2\pi \approx 54.5$ GHz, which is dictated by the trade-off between the hexaferrite magnetic loss and gyrotropy at the operating frequencies.

The junction resonator radius $R \approx 0.99$ mm was initially obtained from the circulation condition (6) at the hexaferrite parameters specified in Table 1. This approximation of R was about 6% larger than the resonator radius R_d deduced from the full-wave simulations, summarized in Table 2. It should be noted that the accuracy of (6) was also limited by the model assumption that the junction resonator is bounded by the magnetic wall at its circumference. Therefore, the junction resonator radius had to be adjusted to account for the effects of fringing fields and discontinuities of the resonator joints with the output transformers. The correction of R was proposed in [24] but the resulting R values were about 30% smaller than those obtained from the full-wave simulations.

Table 2. Dimensions and characteristics of the circulators on hexaferrite substrates.

Junction Resonators	t_s, mm	R_d, mm	a_{tri}, mm	RL *, dB	Iso *, dB	IL *, dB
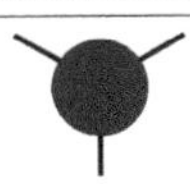	0.17	0.93	-	37.9	28.4	1.33
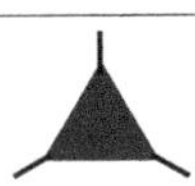	0.09	-	2.46	33.7	21.5	1.66
	0.34	-	2.02	31.5	25.7	1.71

* IL, RL and Iso are at frequency f_0 = 16 GHz.

The full-wave EM simulations in CST MW Studio were adopted here for the analysis and modelling of the circulator performance. The correct value of R was particularly important for the design of the self-biased circulator where the internal magnetic bias could not be adjusted without remagnetizing the entire hexaferrite resonator and/or altering its composition. Thus, the microstrip circulators on hexaferrite substrates were analysed and optimised using the parameters specified in Table 1 and the initial value of R obtained from (6).

3.1. Analysis of Losses in Self-Biased Microstrip Circulators

High ILs are a major problem in the design of self-biased hexaferrite circulators. Recent progress in the synthesis of hexaferrite materials has enabled magnet-free microstrip and waveguide junction circulators with ILs below 1 dB in the frequency band 30–40 GHz [10,11]. However, at lower frequencies, the ILs in the microstrip circulators on hexaferrite substrates remain stubbornly high, exceeding 1.5 dB in the Ku band [8].

In the conventional microstrip circulator with the cross-section shown in Figure 3a, ILs are incurred by both a junction resonator and matching transformers located on the same hexaferrite substrate. Their individual contributions to the total IL depend on the junction resonator shape, and the substrates of different thicknesses are required for the same operating frequency [1,2]. Namely, the substrate of a disk-shaped junction resonator has to be thicker than that of the apex-fed triangular resonator but thinner than the side-fed triangular resonator. The matching transformers have to be adjusted for each substrate thickness, too. Table 2 summarises the simulated IL, return loss (RL) and isolation (Iso) of the disk and triangular-shaped junction circulators with centre frequency f_0 = 16 GHz, and R_d is the radius of the disk-shaped resonator and a_{tri} is the side length of the equilateral triangular resonator. The hexaferrite substrates have the same material parameters specified in Table 1 but different thicknesses t_s.

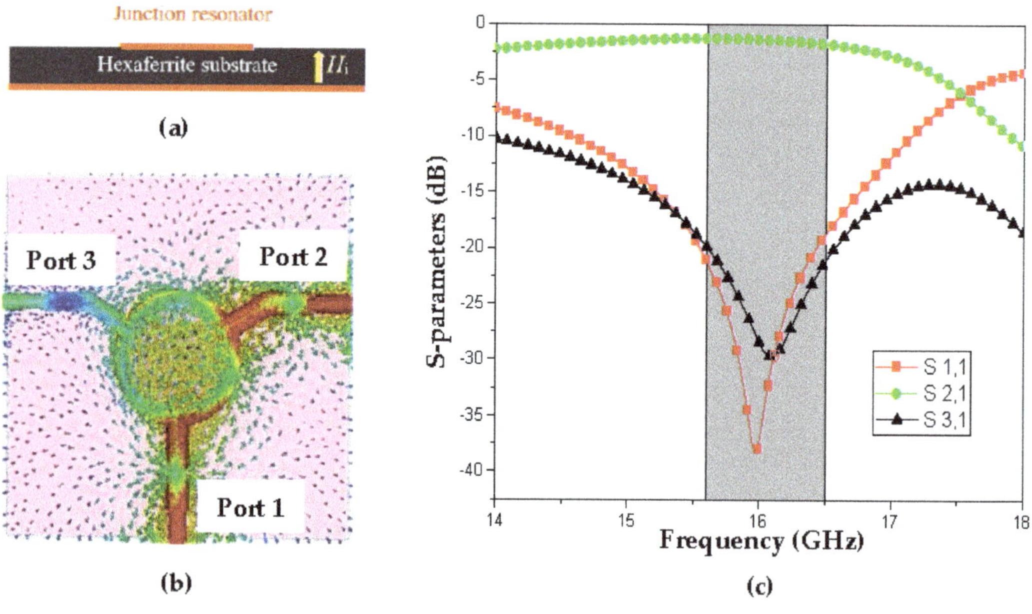

Figure 3. Self-biased disk-shaped junction circulator on hexaferrite substrate of thickness t_s = 0.17 mm biased by the internal magnetic field H_i normal to the surface. (**a**) microstrip circulator cross-section; (**b**) pattern of the RF magnetic field on the substrate top surface at frequency f_0 = 16 GHz; (**c**) the simulated S-parameters.

Table 2 shows that the RL and Iso of all three types of junction resonators are commensurate. But the IL in the circulator with the disk-shaped resonator is lower than ILs in both circulators with triangular resonators. The circulator with the disk-shaped junction resonator on the hexaferrite substrate also exhibits a slightly lower IL than the circulator on the stacked hexaferrite-Duroid substrate reported in [8]. Further optimization of the circulator layout, including the use of tapered matching transformers, noticeably improves the RL and Iso across a 5.6% frequency band, highlighted grey in Figure 3c. But the IL is barely improved and remains high.

The causes of the stubbornly high IL and the effect of hexaferrite substrate thickness t_s on the IL in the disk-shaped junction circulator were examined first. The simulated characteristics of the circulator with several different t_s are summarized in Table 3. They show that as the hexaferrite substrate becomes thinner, FBW broadens whilst the RL and Iso remain high. But the reduction of the IL at smaller t_s is marginal, and the IL decrement becomes even smaller at thinner substrates.

Table 3. Characteristics of the disk-shaped junction circulator on hexaferrite substrates of different thicknesses t_s at centre frequency f_0 = 16 GHz.

t_s, mm	IL, dB	RL, dB	Iso, dB	FBW, %
0.253	1.58	24.27	24.0	3.5
0.17	1.33	39.7	28.4	5.6
0.14	1.31	27.8	35.4	6.7
0.12	1.29	26.0	38.6	6.9

To identify the causes of the circulator's high losses, the RF magnetic field distribution was evaluated on the surface of La-Co-substituted SM-type hexaferrite substrate with a disk-shaped junction resonator and matching transformers. The inspection of the magnetic field pattern in the circulator, Figure 3b, reveals two important features:

- Hot spots exist in the small peripheral regions of the junction resonator near ports 1 and 2 only.
- Standing wave patterns appear in the matching transformers.

A strong magnetic field at the edges of the junction resonator unavoidably increased the dissipative losses. The contribution of the edge to the overall losses was determined by the localisation of the total power flow at the circumference of the junction resonator, which was proportional to the ratio t_s/R_d. Therefore, the IL in the circulators with thinner substrates should be lower, which was confirmed by the simulation results as shown in Table 3.

The field distribution in the matching transformers located on the hexaferrite substrate has the standing wave pattern seen in Figure 3b. Such a pattern causes significant losses that are commensurate or even exceed the losses in the junction resonator itself. To discriminate the individual contributions of the transformers and junction resonator to the total loss, the circulator substrate was modified—the hexaferrite substrate outside the junction resonator disk was replaced by the low-loss alumina substrate of the same thickness as illustrated by Figure 4a. Then the matching transformers were located on the alumina substrate [7]. This modification has several advantages. First, the transformers have no magnetic losses. Second, alumina has a smaller refractive index than hexaferrite puck and increases the field confinement to the junction resonator by reducing the parasitic effect of fringing fields. Thus, the circulator with the composite hexaferrite–alumina substrate has a significantly lower IL as discussed next.

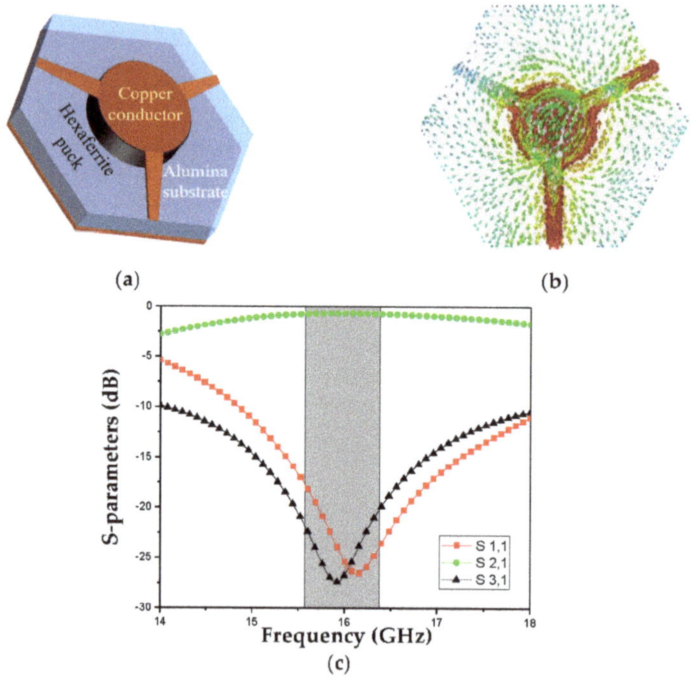

Figure 4. Self-biased disk-shaped junction circulator on a composite hexaferrite–alumina substrate with the internal magnetic bias field H_i normal to the surface. (**a**) microstrip circulator layout; (**b**) RF magnetic field distribution on the substrate top surface at frequency f_0 = 16 GHz; (**c**) simulated S-parameters. The junction resonator radius is 0.91 mm; the transformer length is 1.8 mm and the width of the tapered transformers varies from 0.36 mm at the proximity end to 0.14 mm at the distal end; substrate thickness t_s = 160 μm and hexagon side length is 5.68 mm.

3.2. Self-Biased Microstrip Circulator on Composite Substrate

A circulator with the hexaferrite–alumina substrate of thickness t_s = 0.16 mm and the layout shown in Figure 4a was simulated using the hexaferrite parameters specified in Table 1 and the alumina relative permittivity ε_d = 9.8 and tanδ_d = 0.0001. Even in the basic circulator layout with the quarter-wave matching transformers, the IL at centre frequency f_0 = 16 GHz nearly halved by decreasing from 1.33 dB to 0.79 dB at the RL of 17.5 dB and isolation of 33.2 dB remaining practically unchanged. This is a significant improvement of the IL as compared with the case of the whole circulator on the hexaferrite substrate, illustrated in Table 3. This was the result of the field confinement to the hexaferrite junction resonator and the transformer placement on the low-loss alumina substrate. The FBW of 2.5%, obtained in such a circulator with the composite substrate, remained rather narrow. It was limited by the basic quarter-wave transformers and their suboptimal coupling to the junction resonator. These shortcomings were remedied by the use of the tapered matching transformers that provide a better coupling to the junction resonator.

The circulator layout with tapered impedance transformers on the alumina substrate is shown in Figure 4a. Its IL is further reduced to 0.66 dB at an RL of 25.3 dB and Iso of 26.7 dB at centre frequency f_0 = 16 GHz. The RL over 17 dB and Iso over 20 dB were achieved across the FBW over 5% of the highlighted grey area in Figure 4c. Thus, both IL and FBW are noticeably improved here as compared to the circulator with the quarter-wave transformers. A pattern of the RF magnetic field on the substrate surface is shown in Figure 4b. It demonstrates a fairly smooth wave flow through the junction resonator and matching transformers. Further reduction of the IL can be achieved by optimizing the matching transformers and the hexaferrite with a narrower line of ferrimagnetic resonance ΔH.

The effects of fabrication tolerances and variations of the dimensional and material parameters on the circulator characteristics were assessed. The sensitivity analysis showed that deviations of the junction resonator diameter and internal magnetic field H_i from their nominal values have a stronger effect on the circulator performance than the other parameters. But their effect is limited to a shift of the centre frequency that can be readily adjusted by conditioning the magnetic bias or dimensions of the hexaferrite junction resonator.

4. Self-Biased Microstrip Circulators at Low GHz Frequencies

The analysis of the self-biased circulators in Section 3 was based on the averaged parameters of hexaferrite materials. They were deduced from the measurements whereas the actual distribution and magnitude of the DC magnetic bias in the hexaferrite specimens cannot be obtained directly. Namely, the magnitude and orientation of the internal DC magnetic field inside the polycrystalline hexaferrite layer depend not only on the grain size and density but also on their orientation. In the flat specimens, internal magnetic bias was also inhomogeneous as illustrated by the magnetic field pattern in Figure 2. But the effect of the nonuniform magnetic bias was somewhat mitigated with the aid of the more complicated matching circuits and transformers adjusted for the specified operational frequencies as discussed below.

4.1. Requirements to the Hexaferrite Materials for the Low GHz Self-Biased Circulators

For hexaferrite use at the low GHz frequencies, its internal magnetic bias H_i became smaller than the frequency of $4\pi M_r$. Therefore, the circulators had to be analysed in the two limiting cases of weak and strong internal demagnetisation separately, viz.

- The internal magnetic bias is $H_i = H_a$ — weak demagnetisation (ND).
- The internal magnetic bias is $H_i = H_a - 4\pi M_r$ — strong demagnetisation (SD).

H_i defines the FMR frequency as

$$f_{\mathrm{FMR}} = fl H_i \tag{9}$$

where $\gamma = 2.8$ GHz/kOe is the gyromagnetic ratio and f_{FMR} varies in the wide range

$$fl(H_a - 4\pi M_r) \leq f_{\mathrm{FMR}} \leq fl H_a \tag{10}$$

The geometrical parameters of the junction resonator strongly depend on H_i and differ significantly in the two limiting cases.

The operational frequency of the self-biased circulator had to be away from the f_{FMR} frequency to avoid high losses. But it should not be too far from the f_{FMR} to exploit the hexaferrite gyrotropy. Therefore, the internal magnetic bias H_i of the hexaferrite must be carefully tailored to the desired operational frequency. This causes serious difficulties because H_a should be notably larger than $4\pi M_r$ but small enough for the use at the low GHz frequencies. Therefore, the existing hexaferrite materials developed for the mm-wave frequencies do not meet these requirements.

The losses and the operational band of the circulators are critically dependent on the FMR linewidth ΔH, especially at low GHz frequencies. But synthesis and fabrication of hexaferrites with low H_a and narrow ΔH ($\Delta H \leq 500$ Oe) remain the challenging tasks. The recently developed SM-type hexaferrites have allowed the circulator losses to be reduced below 0.7 dB at the frequencies of 16 GHz as demonstrated in Section 3 but their use at lower frequencies requires further investigations.

4.2. Self-Biased Microstrip Circulators at Low GHz Frequencies

The self-biased microstrip circulators were analysed at frequencies 4 and 6 GHz using the commercially available hexaferrites 08СЧА5В and 08СЧА1В with the parameters are specified in Table 4 [25]. It is necessary to note that the anisotropy field H_a of these hexaferrites is too strong for the considered operational frequencies that results in a weak gyrotropy. Therefore, the designed circulators have narrow bandwidths and high losses at the specified frequencies.

Table 4. Parameters of the hexaferrite materials [25] used for the circulator simulations.

Material	$4\pi M_s$, kOe	$4\pi M_r$, kOe	H_a, kOe	ε_f	tan δ_f	ΔH, Oe
08СЧА5В	3.4	3.06	6	17	0.001	2500
08СЧА1В	3.7	3.33	11	17	0.001	2500

Similar to the circulator discussed in Section 3, a composite substrate was made of a hexaferrite disk embedded in the surrounding alumina layer of the same thickness. The circulator layout with the impedance transformers and advanced matching circuits is shown in Figure 5. The matching circuits were modified for the lower frequency use and to alleviate the limitations imposed by the parameters of the existing hexaferrite materials. Three outstretched lengths of the transmission lines with extension stripes at the ends were added for the impedance matching of the junction. The dimensions of the junction resonators and the matching circuits are summarised in Table 5 for the ND and SD cases.

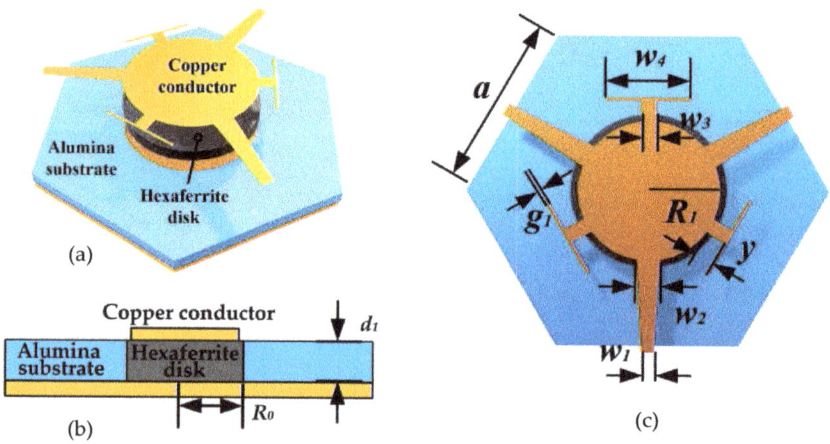

Figure 5. Self-biased microstrip junction circulator with matching circuits: (**a**) exploded view of the circulator layout. (**b**) Cross-section of the structure. (**c**) Top view with the geometrical parameter definitions.

Table 5. Dimensions in mm of the self-biased circulators for 4 GHz and 6 GHz frequency bands.

f_0, GHz	Case	a	w_1	w_2	w_3	w_4	g_1	d_1	y	R_0 *
4	SD	12.0	0.35	1.17	1.1	4.0	0.2	0.4	1.65	2.9
	ND	12.0	0.08	0.62	1.3	4.0	0.2	0.1	1.75	3.6
6	SD	6.0	0.17	0.95	0.6	3.0	0.2	0.2	0.75	2.5
	ND	6.0	0.08	0.93	0.6	2.8	0.2	0.1	1.30	2.6

* The hexaferrite disk radius R_0 is 5% larger than the junction conductor radius R_1.

The S-parameters of the self-biased circulator with the hexaferrite 08СЧА5В were simulated at the centre frequency 4 GHz. Figure 6a,b show the two extreme cases of SD and ND of the hexaferrite disk with internal magnetic bias H_a = 6 kOe and FMR linewidth ΔH = 2.5 kOe. In the SD case, shown in Figure 6a, the isolation exceeds −15dB when frequency varies from 3.92 to 4.08 GHz, i.e., a fractional bandwidth (FBW) is 4%. The RL is above 12.4 dB and the IL varies from 2.8 dB at a centre frequency to 2.9 dB at the band edges. In the ND case shown in Figure 6b, the FMR frequency is further away from the operational band, and the internal magnetic bias is lower. This results in the lower IL ranging from 1.3 dB to 1.9 dB and the RL above 10.7 dB across the specified operational band. But the Isolation band at the level of -20 dB is narrower here, varying from 3.98 GHz to 4.05 GHz.

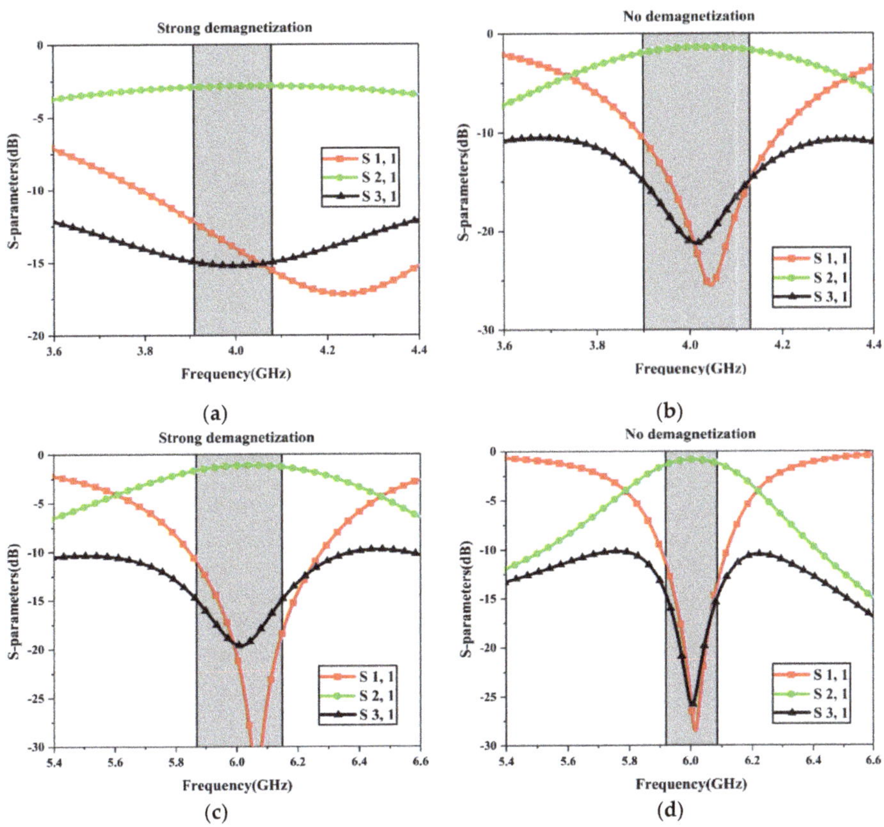

Figure 6. The simulation results of the self-biased microstrip circulators for 4 GHz (**a**,**b**) and 6 GHz (**c**,**d**) in the cases of strong demagnetisation (SD)—(**a**,**c**) and no demagnetisation (ND)—(**b**,**d**).

The characteristics of the self-biased circulator in the 6 GHz band are illustrated in Figure 6c,d. A hexaferrite 08CЧA1B was used here. It has a larger H_a = 11 kOe and the same ΔH = 2.5 kOe. The simulation results in the SD case show that the IL is notably lower, varying from 1.1 dB at the centre frequency to 1.6 dB at the band edges, whilst RL is above 11.2 dB at the band edges. The frequency band of Isolation at the level of −15 dB ranges from 5.87 to 6.15 GHz and the FBW is 4.7%. In the case ND, the IL is lower and varies from 0.8 dB at the centre frequency to 1.3 dB at the band edges, whilst the RL is above 11.6 dB at the band edges. The frequency band of Isolation at the level of -15 dB varies from 5.92 to 6.08 GHz that corresponds to the narrower FBW of 3.7%.

The circulator BW, IL, RL and isolation are the interdependent characteristics. Therefore, their optimal values cannot be realised simultaneously. As a result, some trade-offs can be made in the design of the circulator transformers. The effect of the matching circuit dimensions w_2 and w_4 on the circulator S-parameters is illustrated in Figure 7 for the case of strong demagnetisation. It shows that a higher transmission and wider bandwidth can be obtained at the cost of reducing isolation. By changing w_2 from 0.7 mm to 1.1 mm, the transmission FBW at the level of −3 dB widens from 9.05% to 12.53%, but the IL increases for 0.34 dB and the isolation band becomes narrower and shallower at a larger w_2. A trade-off between the IL and isolation exists when increasing the w_4, which is the length of the matching strips at the end of the outstretched arms shown in Figure 5c. It is seen in Figure 7c that the IL increases from 1.3 dB to 1.36 dB when w_4 increases from 3 mm to 5 mm. But the opposite trend of deepening and widening the isolation curves is observed in Figure 7d. Thus, better matching circuits are necessary for the design of the self-biased circulators in the low GHz frequency bands.

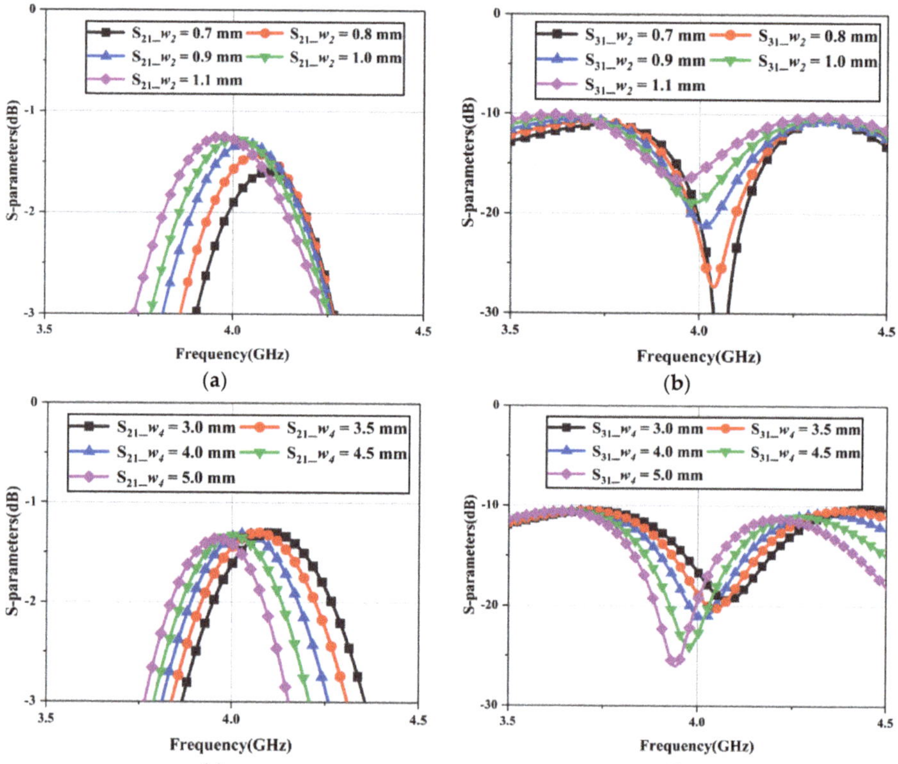

Figure 7. The simulated characteristics of the 4 GHz self-biased microstrip circulator with strong demagnetisation in the cases of variable w_2: (**a**,**b**) and w_4: (**c**,**d**).

5. Conclusions

The mechanisms of RF loss in self-biased microstrip circulators on hexaferrite substrates have been examined, and a means of significant loss reduction in the three frequency bands are discussed. The La-Co-substituted hexaferrites with strong magneto-crystalline anisotropy are used for the design of the self-biased circulators in the Ku, X and C bands.

The pattern of the DC magnetic field is mapped first above the hexaferrite layer surface. It suggests that the magnetic bias is fairly uniform inside a thin hexaferrite layer. The hexaferrite parameters retrieved from the measurement data are used for the full-wave analysis of the self-biased microstrip circulators. It is found that the insertion losses are slightly lower in the circulators with thinner hexaferrite substrates, and this is attributed to a weaker effect of the fringing field at the edges of the junction resonators.

The causes of the high losses in self-biased circulators were examined and individual contributions of the matching transformers and junction resonators were separated. Our analysis has revealed that the matching transformers, located on the hexaferrite substrate, make a major contribution to the circulator's overall losses. To reduce the circulator losses, it is suggested that a composite hexaferrite–alumina substrate be used, where the hexaferrite disk is confined to a junction resonator only whilst the matching transformers are placed on the surrounding alumina substrate. Then the ILs are reduced by more than half. Merits of such an arrangement are illustrated by an example of the Ku band circulator on the hexaferrite–alumina substrate where the IL can be reduced to 0.66 dB at the RL over 17 dB and isolation over 20 dB across the FBW over 5%. These significant improvements of the IL and FBW of the self-biased circulators have made them viable candidates for integration in the RF front-end modules and active antennas.

The self-biased circulators for the low-GHz communications bands have been examined in the C and X bands. In the C band, the 4 GHz self-biased circulators with 08CЧA5B hexaferrite are analysed separately in the extreme cases of the strong and weak demagnetisations. It is found that when the hexaferrite substrate is strongly demagnetised, circulator ILs are high and vary from 2.8 to 2.9 dB at isolation about -15 dB in the FBW of 4%. In the case of weak demagnetization, the IL is lower, varying from 1.3 to 1.9 dB. In the X band, the self-biased circulators are simulated with 08CЧA1B hexaferrite. In the case of strong demagnetisation, the ILs of the circulator are notably lower, varying from 1.1 dB to 1.6 dB across the FBW of 4.7%. In the case of weak demagnetization the operational frequency is farther away from the FMR. This results in the lower ILs varying from 0.8 to 1.3 dB, but the obtained operational bandwidth remains narrow.

The presented analysis demonstrates the great potential of the self-biased circulators for the current and future communications systems operating in the low and mid GHz frequency bands.

Author Contributions: Conceptualization, Y.H. and A.S.; investigation, S.J., L.K. and T.E.; methodology, validation, writing—original draft and visualization, L.K. and S.J.; writing—review and editing, Y.H. and A.S. All authors have read and agreed to the published version of the manuscript.

Funding: This research was funded by Huawei Technologies Sweden AB.

Data Availability Statement: Restrictions apply to the availability of these data. Data is available with the permission of the author.

Acknowledgments: We would like to acknowledge the help and comments from other colleagues at the University of Liverpool and Huawei Technologies Sweden.

Conflicts of Interest: The authors declare no conflict of interest.

References

1. Helszajn, J. *The Stripline Circulator*; John Wiley & Sons: Hoboken, NJ, USA, 2008; ISBN 9780470264140.
2. Linkhart, D.K. *Microwave Circulator Design*; Artech House: Boston, MA, USA, 2014; ISBN 9781608075836.
3. Adams, R.S.; O'Neil, B.; Young, J.L. The Circulator and Antenna as a Single Integrated System. *IEEE Antennas Wirel. Propag. Lett.* **2009**, *8*, 165–168. [CrossRef]
4. Chen, Y.; Geiler, A.L.; Sakai, T.; Yoon, S.D.; Vittoria, C.; Harris, V.G. Microwave and Magnetic Properties of Self-Biased Barium Hexaferrite Screen Printed Thick Films. *J. Appl. Phys.* **2006**, *99*, 08M904. [CrossRef]
5. Geiler, A.L.; Harris, V.G. Atom Magnetism: Ferrite Circulators—Past, Present, and Future. *IEEE Microw. Mag.* **2014**, *15*, 66–72. [CrossRef]
6. Harris, V.G. Modern Microwave Ferrites. *IEEE Trans. Magn.* **2012**, *48*, 1075–1104. [CrossRef]
7. O'Neil, B.K.; Young, J.L. Experimental Investigation of a Self-Biased Microstrip Circulator. *IEEE Trans. Microw. Theory Tech.* **2009**, *57*, 1669–1674. [CrossRef]
8. Wang, J.; Yang, A.; Chen, Y.; Chen, Z.; Geiler, A.; Gillette, S.M.; Harris, V.G.; Vittoria, C. Self-Biased Y-Junction Circulator at Ku Band. *IEEE Microw. Wirel. Compon. Lett.* **2011**, *21*, 292–294. [CrossRef]
9. Peng, B.; Xu, H.; Li, H.; Zhang, W.; Wang, Y.; Zhang, W. Self-Biased Microstrip Junction Circulator Based on Barium Ferrite Thin Films for Monolithic Microwave Integrated Circuits. *IEEE Trans. Magn.* **2011**, *47*, 1674–1677. [CrossRef]

10. Fitchorov, T.; Geiler, A. *Miniaturized, Low-Cost, Self-Biased Circulators for Space and Airborne Applications*; Earth Science Technology Forum: Annapolis, MD, USA, 14–16 June 2016.
11. Laur, V.; Vérissimo, G.; Quéffélec, P.; Farhat, L.A.; Alaaeddine, H.; Laroche, E.; Martin, G.; Lebourgeois, R.; Ganne, J.P. Self-Biased Y-Junction Circulators Using Lanthanum- and Cobalt-Substituted Strontium Hexaferrites. *IEEE Trans. Microw. Theory Tech.* **2015**, *63*, 4376–4381. [CrossRef]
12. Self Biased Circulators and Isolators, Metamagnetics Inc. Available online: https://www.mtmgx.com/rf-microwave-smt-circulators-isolators/ (accessed on 13 November 2022).
13. Kodera, T.; Sounas, D.L.; Caloz, C. Artificial Faraday Rotation Using a Ring Metamaterial Structure without Static Magnetic Field. *Appl. Phys. Lett.* **2011**, *99*, 031114. [CrossRef]
14. Kodera, T.; Sounas, D.L.; Caloz, C. Magnetless Nonreciprocal Metamaterial (MNM) Technology: Application to Microwave Components. *IEEE Trans. Microw. Theory Tech.* **2013**, *61*, 1030–1042. [CrossRef]
15. Reiskarimian, N.; Krishnaswamy, H. Magnetic-Free Non-Reciprocity Based on Staggered Commutation. *Nat. Commun.* **2016**, *7*, 11217. [CrossRef] [PubMed]
16. Sounas, D.L.; Soric, J.; Alù, A. Broadband Passive Isolators Based on Coupled Nonlinear Resonances. *Nat. Electron.* **2018**, *1*, 113–119. [CrossRef]
17. Hessien, M.M.; Radwan, M.; Rashad, M.M. Enhancement of Magnetic Properties for the Barium Hexaferrite Prepared through Ceramic Route. *J. Anal. Appl. Pyrolysis* **2007**, *78*, 282–287. [CrossRef]
18. Drung, D.; Abmann, C.; Beyer, J.; Kirste, A.; Peters, M.; Ruede, F.; Schurig, T. Highly Sensitive and Easy-To-Use SQUID Sensors. *IEEE Trans. Appl. Supercond.* **2007**, *17*, 699–704. [CrossRef]
19. Labeyrie, M.; Mage, J.; Robinson, T. Characterisation of Strontium Hexaferrite for Millimeter Waves Applications. *IEEE Trans. Magn.* **1986**, *22*, 976–980. [CrossRef]
20. Baker-Jarvis, J.; Vanzura, E.J.; Kissick, W.A. Improved Technique for Determining Complex Permittivity with the Transmission/Reflection Method. *IEEE Trans. Microw. Theory Tech.* **1990**, *38*, 1096–1103. [CrossRef]
21. Lax, B.; Button, K.J. *Microwave Ferrites and Ferrimagnetics*; McGrow-Hill: New York, NY, USA, 1962.
22. Bosma, H. On Stripline Y-Circulation at UHF. *IEEE Trans. Microw. Theory Tech.* **1964**, *12*, 61–72. [CrossRef]
23. Fay, C.E.; Comstock, R.L. Operation of the Ferrite Junction Circulator. *IEEE Trans. Microw. Theory Tech.* **1965**, *13*, 15–27. [CrossRef]
24. Wu, Y.S.; Rosenbaum, F.J. Wide-Band Operation of Microstrip Circulators. *IEEE Trans. Microw. Theory Tech.* **1974**, *22*, 849–856. [CrossRef]
25. Hexaferrite Materials. Available online: https://www.domen.ru/files/upload/2020-12-24_15:46:22_25d13f31ce5136b6.pdf (accessed on 12 November 2022). (In Russian).

magnetism MDPI

Article

Numerically Stable and Computationally Efficient Expression for the Magnetic Field of a Current Loop

Michael Ortner *, Peter Leitner and Florian Slanovc

Magnetic Microsystem Technologies, Silicon Austria Labs, 9500 Villach, Austria
* Correspondence: michael.ortner@silicon-austria.com

Abstract: In this work, it is demonstrated that straightforward implementations of the well-known textbook expressions of the off-axis magnetic field of a current loop are numerically unstable in a large region of interest. Specifically, close to the axis of symmetry and at large distances from the loop, complete loss of accuracy happens surprisingly fast. The origin of the instability is catastrophic numerical cancellation, which cannot be avoided with algebraic transformations. All exact expressions found in the literature exhibit similar instabilities. We propose a novel exact analytic expression, based on Bulirsch's complete elliptic integral, which is numerically stable (15–16 significant figures in 64 bit floating point arithmetic) everywhere. Several field approximation methods (dipole, Taylor expansions, Binomial series) are studied in comparison with respect to accuracy, numerical stability and computation performance. In addition to its accuracy and global validity, the proposed method outperforms the classical solution, and even most approximation schemes in terms of computational efficiency.

Keywords: magnetic field; current loop; analytic solution; numerical stability; computation performance

1. Introduction and Motivation

Analytic expressions for the magnetic field of current and magnetization problems are widely used in modern science and engineering. They offer much faster field computation than their numerical counterparts, and the superposition principle often makes up for the simple geometries for which solutions exist. User friendly, but much slower numerical alternatives like ANSYS Maxwell [1] or Comsol [2] are often not available due to their substantial financial costs, and implementations through open-source packages like FEniCS [3], or NG-Solve [4] are time consuming and require substantial know-how.

In addition, analytic forms enable extreme computation precision up to 12–14 significant digits in Double (64 bit) floating point arithmetic when they are properly implemented. The downside is that the expressions found in the literature are mostly arranged to look as simple as possible, with little or no thought given to their implementation. A straightforward transfer to the computer code often results in numerical instabilities close to symmetry positions and special cases, surface and edge extensions, and at large distances from the magnetic field sources.

A good example is the classical textbook expression for the radial component of the B-field of a circular current loop [5],

$$B_\rho = \frac{\mu_0 i_0}{2\pi} \frac{z}{((\rho+\rho_0)^2+z^2)^{1/2}} \left[-K(k^2) + \frac{\rho^2+\rho_0^2+z^2}{(\rho-\rho_0)^2+z^2} E(k^2) \right], \tag{1}$$
$$\text{with} \quad k^2 = \frac{4\rho_0\rho}{(\rho+\rho_0)^2+z^2}.$$

The current loop with radius ρ_0 lies in the $z=0$ plane, with center in the origin of a cylindrical coordinate system (ρ, φ, z), and carries a current i_0. The vacuum permeability is denoted by μ_0, and K and E are the complete elliptic integrals of first and second kind, respectively, for which fast and stable numerical algorithms exist [6,7] and implementations are provided by common tools like Mathematica [8], Matlab [9] and Scipy [10].

A straightforward implementation $\hat{B}_\rho$ of expression (1) is numerically troublesome because catastrophic cancellation happens close to the axis, $\rho \ll \rho_0$, and at large distances from the loop, $\rho^2 + z^2 \gg \rho_0^2$. The magnitude of this phenomenon is easily underestimated. For example, evaluations of $\hat{B}_\rho$ in Double floating point format give only four correct significant figures when evaluated at not very distant positions $\boldsymbol{r} = (0.01\rho_0, \varphi, 100\rho_0)$. The extent of the problem is laid out in Figure 1. There, the first quadrant in ρz-space is depicted for an arbitrary φ. The other quadrants follow from the symmetry of the problem.

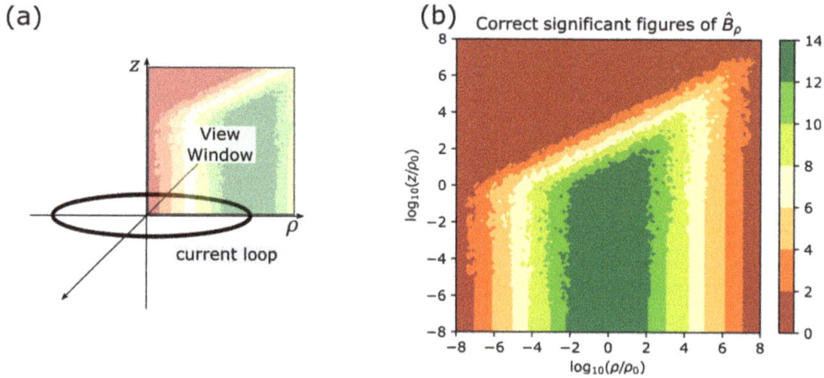

Figure 1. (**a**) Sketch of a current loop and the observed first quadrant. (**b**) The number of correct significant figures of a straightforward implementation of the textbook expression for the radial component of the B-field of a current loop on a log–log scale.

Discussions in the Magpylib [11] forum on Github underline the practical relevance of the issue. Magpylib is an open-source Python package that combines analytic solutions of permanent magnets and current problems with a geometric reference frame API. It enables users to easily compute the magnetic fields of sources at observers with arbitrary relative position and orientation. In the now outdated version 3.0.5, a numerically unstable, straightforward implementation is used for the current loop field, together with a special case for $\rho = 0$. When an unsuspecting user creates such a current loop object and rotates it about an observer located on the loop axis, see Figure 2a, the limited precision of the rotation naturally results in a misalignment between the observer and loop axis by small values of the order of the machine precision. In the reference frame of the current loop, the observer is thus moved from the stable special case, $\rho = 0$, to the numerically unstable position $0 < \rho \ll \rho_0$. The resulting bad computation is shown in Figure 2b for multiple angles α ranging from 0 to 360 degrees. Complete loss of precision can be observed in many cases.

Achieving numerical stability is of critical importance. Instabilities like the one described in this work are often not visible, but can lead to erroneous computation results that are difficult to track down. In this paper, we analyze various expressions for the B-field of a current loop in terms of range of validity, precision, numerical stability and performance. Eventually, a reader can decide by himself which formulation suits his needs best.

The structure of the manuscript is as follows: In Section 2, the classical textbook expression is discussed, the origin of the problem is identified, and several expressions for the B-field are introduced and analyzed in terms of accuracy. In Section 3, all these expressions are tested in terms of computation performance. In Section 4, all results are reviewed and discussed. Finally, in Appendix A, various algorithms used in this work are provided.

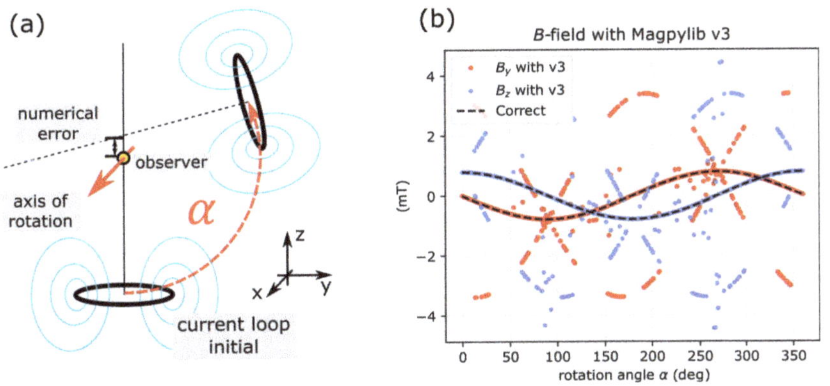

Figure 2. (**a**) Sketch of typical current loop positioning with the package Magpylib. (**b**) Demonstrating the relevance of the numerical instability, which becomes visible when a current loop rotates about an observer.

2. Numerical Stability

2.1. Fundamentals and Annotation

As this paper is intended mostly for physicists and engineers, we outline a few basic concepts and annotation details common in numerical analysis, following the textbook *Accuracy and Stability of Numerical Algorithms* [12].

Evaluation of an expression in floating point arithmetic is denoted $fl(\cdot)$. A basic arithmetic operation op $\in \{+,-,*,/\}$ satisfies

$$fl(x \text{ op } y) = (x \text{ op } y)(1+\delta), \quad |\delta| \leq u. \tag{2}$$

Here, u is the machine precision, which is $2^{-52} \approx 1.11 \times 10^{-16}$ for the standard 64 bit Double floating point format, which is also used in this work. In contrast to the textbook, we use the common notation that bold symbols denote vectors, and also do not distinguish between accuracy and precision.

Computed quantities wear a hat, i.e., $\hat{f}$ denotes the computed value of an expression f. The absolute error of a computed quantity is defined as, $E_{\text{abs}}(\hat{f}) = |f - \hat{f}|$. In relation to the function value, it gives the important relative error

$$E_{\text{rel}}(\hat{f}) = \frac{|f-\hat{f}|}{|f|}. \tag{3}$$

The intuitive measure of *correct significant figures* of an evaluation is a flawed concept, as explained in the textbook, but is used here nonetheless by rounding the relative error. In most cases, an evaluation of $\hat{f}$ experiences large relative errors about zero crossings of the function f, accompanied by a respective loss of significant figures. However, this is generally not a problem as long as the absolute error is reasonably bounded there, $E_{\text{abs}}(\hat{f}) \ll \epsilon$.

We speak of *numerical stability* when the relative error is small, or at zero crossings, when the absolute error is, respectively, smaller than the mean absolute function value in the vicinity of the crossing. In Double floating point format, proper implementations of analytic expressions are expected to give relative errors below 10^{-10} to 10^{-12}, or in the vicinity of zero-crossings, absolute errors that are at least 10–12 orders of magnitude smaller than the mean absolute function value.

When a vector field g is studied, this is done componentwise with (3), or vectorwise using the Euclidean norm $||\cdot||$,

$$E_{\text{rel}}(\hat{g}) = \frac{||g - \hat{g}||}{||g||}. \tag{4}$$

Such a measure can also mitigate the apparent relative error problem about zero-crossings of individual components.

Rounding happens when an internal operation results in a number that has more decimal places than is provided by the used floating point arithmetic. For example, the input $a = 8$ can be fully represented in Single (32 bit) precision, while the result of $b = a/3 = 2.6666667$ is rounded off. Rounding is by itself not a problem, since the error appears only in the last digit. The relative error according to (3) yields $E_{\text{rel}}(\hat{b}) = 1.25 \times 10^{-8}$, which is in the expected range for Single precision with 7–8 significant digits. However, this rounding error sets the stage for *numerical cancellation*. Consider that a number $c = 2.6666$ is subtracted from the previous result. The outcome is $d = b - c = 6.67 \times 10^{-5}$ and has only three significant figures. The resulting relative error increases to $E_{\text{rel}}(\hat{d}) = 5 \times 10^{-4}$ in this case. Further computation with this value must be treated with great care. Numerical cancellation happens when subtracting rounded values of equal size from each other, and can even lead to a complete loss of precision, i.e., no correct significant figures.

A power series can be assumed to be stable when implementing the sum with falling order (small terms first). The limit of precision of a series expansion is mostly a result of the chosen truncation.

In this work, all quantities are evaluated in Python v3.9.9, and making use of the standard packages Numpy v1.21.5 and Scipy v1.7.3. Many of the used implementations are reproduced in Appendix A.

2.2. The Classical Solution

The exact analytic expressions for the off-axis magnetic field of a circular current loop have been known for a long time. They can either be derived by direct integration over the current density, or by solving a boundary value problem. A few examples included in classical textbooks show expressions derived in cylindrical coordinates [5] and in spherical coordinates [13] via the vector potential. A solution via the boundary value problem is demonstrated in [14]. A detailed summary of analytic expressions for the current loop field and their derivatives in different coordinate systems is given in [15]. More recent developments show a computation through the law of Biot–Savart where the result is expressed in terms of hypergeometric functions [16]. In [17], the field of arc segments is reviewed and expressed in terms of incomplete elliptic integrals, where the special case of a complete current loop is shown to yield the classical results.

Two recent works focus on computationally efficient approximation schemes: Prantner et al. [18] use local Taylor series approximations about supporting points. This has the advantage that the resulting expressions are simple, numerically stable and fast to compute. On the other hand, the validity is limited to small regions about the supporting points, and for each new region, a new series must be constructed. A very elegant solution was proposed by Chapman et al. [19]. They make use of a binomial expansion and reduce the field to a simple series with a high level of accuracy everywhere. Their work aligns strongly with this paper, is reproduced in Section 2.6, and analyzed in terms of performance in Section 3.

Hints towards stability problems of the classical solutions are found in [20], where the authors use a series expansion to "obtain simpler expressions valid at intermediate distances", and in [21], where various analytic representations are derived and studied. Proper numerical implementation, numerical stability and computation performance are hardly discussed in the literature beyond basic algorithms for the complete elliptic integrals and Taylor series expansions.

All works that offer exact analytic forms for the field of the current loop end up with the same expressions in cylindrical coordinates, or with expressions that are easily transformed to the following one

$$B_\rho^{\text{Classic}} = B_0 \frac{k}{\sqrt{\bar\rho}} \frac{\bar z}{\bar\rho} \frac{1}{q^2} \underbrace{\left[(1+q^2)E(k^2) - 2q^2 K(k^2)\right]}_{\xi_0(k,q)}, \tag{5a}$$

$$B_z^{\text{Classic}} = B_0 \frac{k}{\sqrt{\bar\rho}} \frac{1}{q^2} \underbrace{\left[2q^2 K(k^2) - \left(1 + q^2 - \frac{k^2}{\bar\rho}\right) E(k^2)\right]}_{\xi_1(\bar\rho,k,q)} \tag{5b}$$

$$\text{with} \quad k^2 = \frac{4\bar\rho}{\bar z^2 + (1+\bar\rho)^2}, \qquad q^2 = \frac{\bar z^2 + (1-\bar\rho)^2}{\bar z^2 + (1+\bar\rho)^2}.$$

The loop lies in the $z = 0$ plane of a cylindrical coordinate system (ρ, φ, z), with the origin at its center. B_ρ and B_z denote radial and axial component of the B-field, respectively. Everything is expressed in dimensionless quantities, $\bar\rho = \rho/\rho_0$ and $\bar z = z/\rho_0$ with the loop radius ρ_0, and $B_0 = \mu_0 i_0 / 8\pi\rho_0$ with the vacuum permeability μ_0. The functions K and E denote the complete elliptic integrals of first and second kind, respectively, for which fast and stable algorithms exist [6,10].

The formulation is chosen through the two dimensionless quantities k and q that satisfy

$$q^2 = 1 - k^2. \tag{6}$$

While k and q can be expressed through each other, it is important that they are computed individually to avoid numerical cancellation effects. Specifically, a pure k-formulation results in additional problems about the singularity at $\bar\rho = 1, \bar z = 0$, while a pure q-formulation adds to the main instability described in the next section.

The system is independent of the azimuth angle φ, so that the parameter space of interest, covering all of $\mathbb{R}^3$ except the loop itself, is,

$$\{\bar\rho \in [0,\infty), \bar z \in (-\infty,\infty) : \bar\rho \neq 1 \vee \bar z \neq 0\} \quad \rightarrow \quad k, q \in [0,1]. \tag{7}$$

Due to the symmetry, we only study positive values of $\bar z$.

2.3. Numerical Stability of the Classical Solution

By straightforward implementation, we mean naive transfer of a function to the computer script. A code example for such an implementation $\hat{\boldsymbol{B}}^{\text{Classic}}$ of Equation (5) is provided in Appendix A.1. This implementation is component- and vectorwise numerically unstable, which can be seen from the relative errors shown in Figure 3. An increasing loss of precision with distance from the loop ($\bar\rho^2 + \bar z^2 \gg 1$) can be observed in all subplots. In addition, the radial component becomes unstable towards the axis ($\bar\rho \ll 1$). This instability translates partially to the vectorwise error in (c), despite the dominating amplitude of the stable B_z^{Classic}. The troublesome "speckled" region in (c) is the result of the competing error between the two components. Finally, the relative error diverges naturally in a narrow band about the zero crossing of the z-component, outlined with a dotted line in (b). All error computations in this work are achieved by comparison to stable forms that are derived below.

The origin of the observed instabilities is numerical cancellation when evaluating the functions ξ_0 and ξ_1 in (5). Both functions are made up of two summands that tend to π and $-\pi$, respectively, for $k \to 0$, which corresponds to regions close to the axis and far from the loop, as can be seen from Figure 3d.

The cancellation effect is best demonstrated with the function ξ_0. Figure 4a shows a straightforward implementation of ξ_0, and an implementation of its Taylor series for small k including eight terms up to order k^{18}. The fluctuation of $\hat\xi_0$ with decreasing k is a result of the numerical cancellation, which ends in a complete loss of accuracy below $k^2 \approx 3.5 \times 10^{-8}$. Loss of precision resulting from the truncation of the Taylor expansion, as k approaches 1, is made visible in the inset figure.

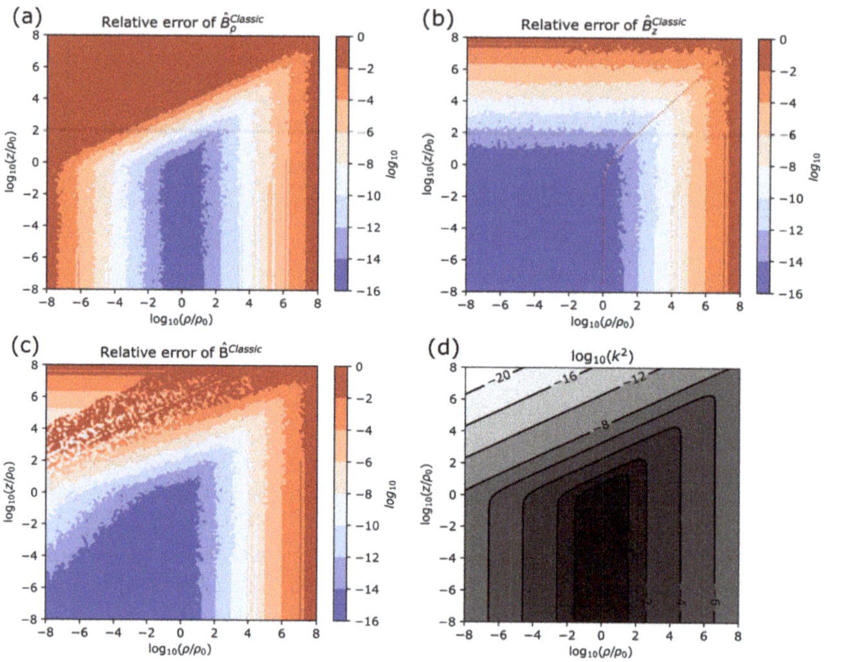

Figure 3. Relative error of a straightforward implementation of the textbook expressions for the B-field of a current loop. (**a**) Radial and (**b**) axial components, as well as (**c**) vectorwise analysis reveal a high level of numerical instability. (**d**) The relation between the cylindrical coordinates and the important quantity k.

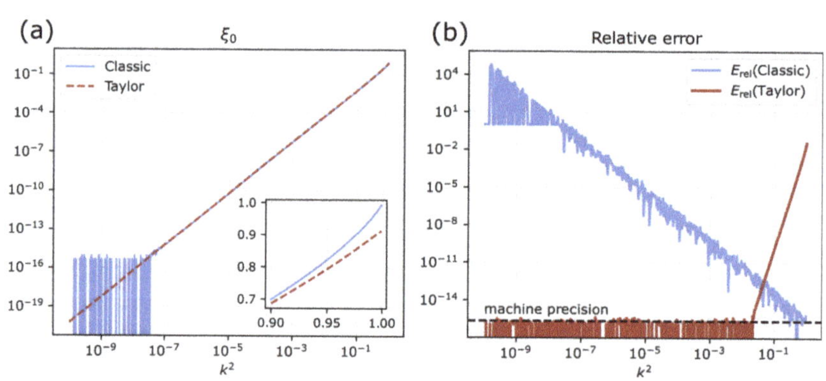

Figure 4. (**a**) Two different implementations of the function ξ_0 and (**b**) the relative difference when comparing them to each other.

Relative errors of the two implementations are shown in Figure 4b. With decreasing k the straightforward implementation loses precision. Below $k^2 \approx 3.5 \times 10^{-8}$ the relative error exceeds 1 which is understood as a complete loss of precision. On the other hand, the accuracy of the Taylor series increases with decreasing k until the truncation error undercuts the machine precision. For small values of k, this Taylor series can be used as an excellent numerical reference.

All exact expressions for the magnetic field of a current loop that are provided in the literature seem to exhibit similar instabilities. This includes the tempting form in spherical coordinates [13,15]. Rearrangement of the problematic summands can partially reduce some cancellation effects, but the main problem remains.

2.4. Dipole Approximation

Dipole models are commonly used to describe the magnetic field at large distances from the sources. They are numerically stable, easy to implement and fast to compute. Specifically for current loops, dipole models are used when coupling distant coils [22], or in some geomagnetic works [23,24]. The relation between the magnetic dipole moment and the current loop parameters is $\boldsymbol{m} = (0, 0, i_0\rho_0^2\pi)$. The resulting dipole field in cylindrical coordinates is given by

$$\boldsymbol{B}^{\text{Dipole}} = B_0 \frac{2\pi}{(\bar{\rho}^2 + \bar{z}^2)^{5/2}} \begin{pmatrix} 3\bar{\rho}\bar{z} \\ 0 \\ 2\bar{z}^2 - \bar{\rho}^2 \end{pmatrix}. \tag{8}$$

Straightforward implementations of (8) are numerically stable, so that the precision of the dipole model is limited only by the approximation error. The vectorwise relative error is shown in Figure 5a. It can be observed that the accuracy of the dipole approximation increases only slowly with distance from the loop. At $\bar{\rho} = 100$, the relative error is still above 10^{-4}. The large difference between current loop and dipole model is also noted in [20].

The colorful contour lines in the figure correspond to the vectorwise relative error of $\hat{\boldsymbol{B}}^{\text{Classic}}$, and are taken from Figure 3c. While the dipole approximation saves the day at large distances, the figure also reveals that both computations fail to give correct results in a large region of interest close to the axis.

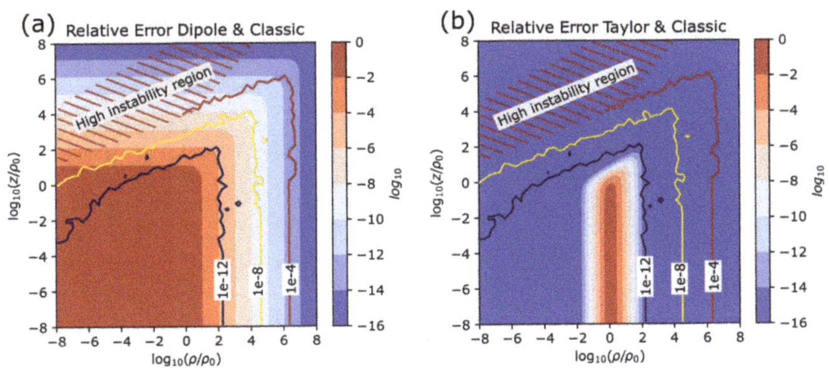

Figure 5. Vectorwise relative error of the dipole approximation (**a**) and a Taylor approximation (**b**) of the B-field of a current loop. The colored contour lines show the numerical error from a straightforward implementation of the classical textbook expressions.

2.5. Taylor Approximation

In Section 2.3, it is explained that the classical textbook expressions suffer from the numerical instabilities of $\xi_0(k,q)$ and $\xi_1(\bar{\rho},k,q)$ for small values of k. A series expansion seems like a natural choice. Taylor approximations of (5a) and (5b) about $k=0$ calculate as

$$B_\rho^{\text{Taylor-k}} = B_0 \frac{\bar{z}}{\bar{\rho}^{3/2}} \frac{3\pi}{16q^2}\left[k^5 + \frac{1}{4}k^7 + \frac{15}{128}k^9 + \mathcal{O}(k^{11})\right], \tag{9a}$$

$$B_z^{\text{Taylor-k}} = B_0 \frac{1}{\bar{\rho}^{3/2}} \frac{\pi}{2q^2}\left[k^3 - \frac{2+3\bar{\rho}}{8}k^5 - 3\frac{1+3\bar{\rho}}{64}k^7 + \mathcal{O}(k^9)\right]. \tag{9b}$$

The vectorwise error of such a series expansion with up to order k^{21} (10 terms) is shown in Figure 5b. Such an expansion with a simple cutoff criterion $k_c = 0.26$ can be combined with the classical solution (5) to yield at least 10 significant digits everywhere.

Moreover, it is possible to provide a second Taylor series about $q=0$, which is accurate when the k-series fails,

$$B_\rho^{\text{Taylor-q}} = B_0 \frac{k\bar{z}}{\bar{\rho}^{3/2}}\left[\frac{1}{q^2} + \frac{3}{4}\ln\left(\frac{eq^2}{16}\right) + \frac{3}{64}\ln\left(\frac{e256}{q^4}\right)q^2 + \mathcal{O}(q^4)\right], \tag{10a}$$

$$B_z^{\text{Taylor-q}} = B_0 \frac{k}{\bar{\rho}^{3/2}}\left[(1-\bar{\rho})\frac{1}{q^2} - \frac{1}{4}\left(4 + \ln\left(\frac{eq^2}{16}\right)(1+3\bar{\rho})\right) + \mathcal{O}(q^2)\right]. \tag{10b}$$

Here e denotes Euler's number. It is shown in Figure 6 that it is possible to cover all of $\mathbb{R}^3$ by overlapping a k-series and a q-series. Specifically, two such series that include terms up to order q^{60} and k^{61} with cutoff criterion $k_c = 0.7$ globally undercut a componentwise relative error of 10^{-10}. The respective $k_c = 0.7$ contour is shown in Figure 6a,b.

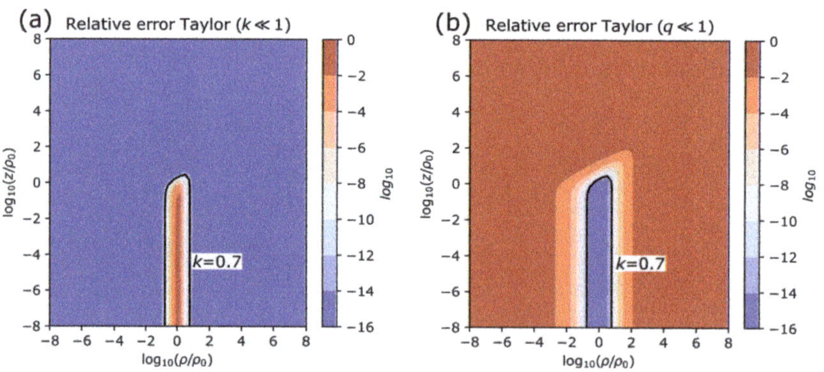

Figure 6. Vectorwise relative error of Taylor approximations of the B-field of a current loop. (**a**) Expansion for small k. (**b**) Expansion for small q.

Proper implementations of these series are provided in Appendix A.2, together with detailed information about the series precision with respect to the chosen truncation. Computation performance is discussed in Section 3.

2.6. Binomial Expansion

It is proposed in [19] to make use of a binomial expansion to generate the following series approximation for the B-field of a current loop,

$$B_\rho^{\text{Binom}} = B_0 \frac{2\pi\bar{z}\bar{\rho}}{(\bar{\rho}^2+\bar{z}^2+1)^{5/2}} \sum_i A_i F_i, \tag{11a}$$

$$B_z^{\text{Binom}} = B_0 \frac{2\pi\bar{\rho}}{(\bar{\rho}^2+\bar{z}^2+1)^{5/2}} \sum_i \left(\frac{4}{W^{1/2}} B_i - \bar{\rho} C_i\right) F_i. \tag{11b}$$

The quantity W is closely related to k^4,

$$W = \frac{4\bar{\rho}^2}{(\bar{\rho}^2+\bar{z}^2+1)^2}, \tag{12}$$

and the functions F_i and coefficients A_i for the lowest five orders are given by

$$\begin{aligned}
F_0 &= 1, & A_0 &= 3,\\
F_1 &= W/(1-W), & A_1 &= 3.601265264628424,\\
F_2 &= (\ln(1-W)+W)/W, & A_2 &= 7.270215646065023\times10^{-1},\\
F_3 &= \ln(1-W), & A_3 &= -5.22255035327797\times10^{-2},\\
F_4 &= W, & A_4 &= -8.69531084533186\times10^{-3},\\
F_5 &= W^2, & A_5 &= -1.4574683941872\times10^{-3}.
\end{aligned}$$

The corresponding coefficients B_i and C_i are

$$\begin{aligned}
B_0 &= 1, & C_0 &= 3,\\
B_1 &= 9.003162495383443\times10^{-1}, & C_1 &= 3.601264998153377,\\
B_2 &= -3.044435633189330\times10^{-2}, & C_2 &= 7.276276696916009\times10^{-1},\\
B_3 &= -2.5946696478408\times10^{-2}, & C_3 &= -5.27599282256922\times10^{-2},\\
B_4 &= -3.9595895609185077\times10^{-3}, & C_4 &= -8.92318142802500\times10^{-3},\\
B_5 &= -7.177536438\times10^{-4}, & C_5 &= -1.5500658123898\times10^{-3}.
\end{aligned}$$

Here, we have simply reproduced the results from Chapman et al. in a format that aligns with this work. Coefficient C_3 has a different sign due to a typo in the original publication.

When implementing the above expressions, great care must be taken with the functions F_2 and F_3, as they are numerically unstable for small values of W. For this purpose, special implementations, commonly termed $log1p$ and $log1pmx$, can be used. The resulting accuracy is displayed in Figure 7, and confirms the claims made in [19], that more than five significant digits are achieved everywhere with this approximation.

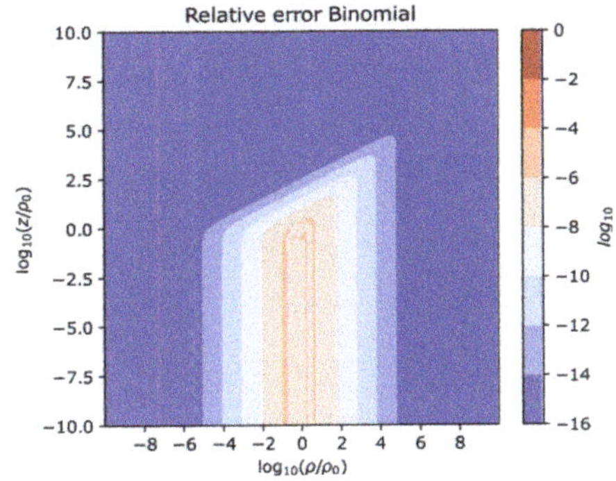

Figure 7. Vectorwise relative error of the binomial expansion for the B-field of a current loop.

2.7. An Exact and Stable Representation

Bulirsch's general complete elliptic integral [25] is defined as

$$cel(k_c, p, c, s) = \int_0^{\pi/2} \frac{c\cos^2\varphi + s\sin^2\varphi}{(\cos^2\varphi + p\sin^2\varphi)\sqrt{\cos^2\varphi + k_c^2\sin^2\varphi}} d\varphi. \tag{13}$$

An efficient algorithm exists for the computation of *cel*. Its original implementation by Bulirsch [25] is reviewed in [26] and is reproduced here in Appendix A.3. A slightly faster, but more verbose implementation by Fukushima can be found in [27]. Algorithm convergence and performance is discussed in Section 3.

It is pointed out in [28] that sums of complete elliptic integrals can be expressed in terms of *cel*,

$$\lambda K(m) + \mu E(m) = cel(\sqrt{1-m}, 1, \lambda + \mu, \lambda + \mu(1-m)). \tag{14}$$

This identity enables us to re-express the troublesome functions ξ_0 and ξ_1 in the form of

$$\xi_0(k, q) = cel(q, 1, k^2, -k^2q^2), \tag{15}$$

$$\xi_1(\bar{\rho}, k, q) = cel(q, 1, k^2(1 - 1/\bar{\rho}), -k^2q^2(1 + 1/\bar{\rho})). \tag{16}$$

Again, we have chosen a favorable argument representation through k and q to avoid instabilities about the singularity at $\bar{z} = 0$ and $\bar{\rho} = 1$.

In principle, the transformation from E and K to *cel* eliminates the numerical cancellation problems of ξ_0 and ξ_1. However, the implementations by Bulirsch and by Fukushima become themselves numerically unstable for small values of k when evaluating (15) and (16).

A closer look at the algorithm in Appendix A.3 reveals the problem: At first (line 7–34), a set of parameters is prepared from the input arguments, thereon (line 35–44) the solution is computed iteratively. For the input arguments in (15), it happens that line 28 becomes $cc = k^2 - k^2q^2$, which exhibits cancellation for $k \ll 1$. This problem is easily fixed by rewriting the expression as $cc = k^4$ in that line. Similar problems, which can be solved by algebraic transformations in the first part of the algorithm, happen also when computing ξ_1 in (16).

In Appendixes A.4 and A.5, we show modified *cel* algorithms, termed $cel^*(k, q)$ and $cel^{**}(\bar{\rho}, k, q)$ for computing ξ_0 and ξ_1, respectively, that are stable and fast. For improved performance, the first part of Bulirsch's original algorithm is shortened down by assigning all parameters their respective values, and the improvements proposed by Fukushima [27] are adapted. Comparisons to overlapping Taylor series expansions, introduced in Section 2.5 confirm the global numerical stability of the two algorithms.

As all other terms in (5) are numerically stable, the following expressions will give the components of the B-field of a current loop with machine precision everywhere,

$$B_\rho^{\text{Stable}} = B_0 \frac{\bar{z}}{\bar{\rho}} \frac{k}{q^2\sqrt{\bar{\rho}}} cel^*(k, q), \tag{17a}$$

$$B_z^{\text{Stable}} = B_0 \frac{k}{q^2\sqrt{\bar{\rho}}} cel^{**}(\bar{\rho}, k, q). \tag{17b}$$

2.8. Loss of Precision at Sign Change

The z-component of the B-field exhibits a sign change when passing from small to large values of $\bar{\rho}$. In (17b), the sign change is revealed by the following relation:

$$cel^{**}(\bar{\rho}, k, q) = \frac{k^2}{\bar{\rho}} E(k^2) - cel^*(k, q). \tag{18}$$

Both functions E and cel^* are positive. Towards the coordinate axis, the first term dominates, while for large values of $\bar{\rho}$, the second term does. The zero-crossing in-between corresponds to the change in sign.

The relative error of the z-component naturally diverges around the zero-crossing. As explained in Section 2.1, we do not speak of numerical instability in this case, however, a loss of significant digits of the z-component still occurs. How fast the loss of significant digits occurs depends on the gradient of the field. In this case, a helpful criterion is found empirically

$$\text{lost digits} \approx \log_{10} |B_\rho / 2B_z|, \tag{19}$$

which corresponds only to a very narrow band about the crossing.

3. Performance

Analytic solutions are mostly used for their computation efficiency, so that the performance of the proposed algorithms is critical. For a performance comparison on equal footing, all algorithms are implemented for scalar inputs in native Python, and making use of the standard *math* package. This means that only sequential execution is considered, the potential for code vectorization is not analyzed here. Algorithms for the complete elliptic integrals, that can also be found in the Scipy library, have been reproduced. All tests were run on CentOS Linux 7.9.2009 operating system on a AMD EPYC 7513 2.6 GHz CPU architecture with 32 cores.

The performance test is designed as follows: Each B-field expression is evaluated for input parameters $(\bar{\rho}_0, \bar{z}_0)$. This is repeated for a set of 200 different points, that all correspond to a similar value of k. The points are equally distributed on the respective k^2 contour lines that are shown in Figure 3d. This procedure is repeated in a loop 1000 times. For each value of k, every expression is thus evaluated 200,000 times. The order of evaluation is mixed up by design, to avoid distortions from computation performance spikes resulting from Python garbage collection and hardware issues. For each k, the computation times of every tested expression are summed up.

The results of this test are shown in Figure 8a for different values of $k^2 \in [10^{-12}, 1)$. Computation times are displayed with respect to the fastest one, which is always the dipole expression. Most algorithms evaluate the same expression independent of k, and show similar timings for all inputs. The proposed stable algorithm is of an iterative nature, and requires less iterations for smaller values of k. The steps from four iterations close to the loop filament down to zero iterations are clearly visible in the figure. The binomial approximation exhibits a step at about $k^2 \approx 0.2$, which results from the special implementation of $log1pmx$ that was used here. This function is not available in Python *math*, and makes use of a 4-term power series for small W.

In Figure 8b, we show the maximal vectorwise relative computation errors of the different methods. Interestingly, convergence and divergence of most methods can be observed well as functions of k^2. For the dipole approximation only positions with $\bar{\rho} \geq 1$ on the k^2 contour lines are included, because it does not converge close to the origin. The stable solution is not visible in this figure, because it is used as a reference. It has machine precision everywhere.

The combination of the two figures enables users to choose the scheme that is most suitable for their purpose. When only computation time and accuracy are taken as measures, Figure 8 shows that the proposed method is extremely competitive. The only solutions that are faster and have similar precision are the Taylor k-series and the dipole approximation, both with limited validity. However, in many cases users might not require machine precision, and can be compelled by the simple analytic expressions offered by the approximation schemes, which also enable effortless computation of the derivatives.

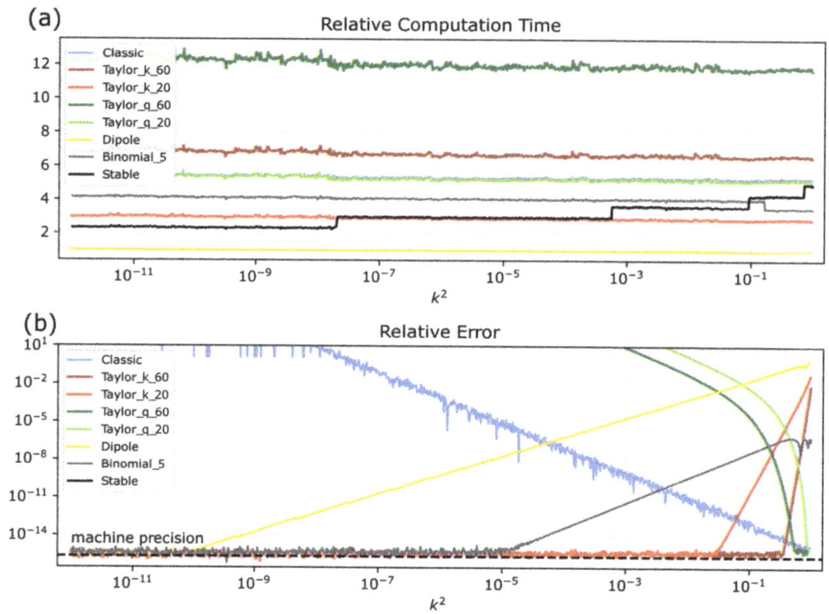

Figure 8. (**a**) Computation times of various methods with respect to the fastest one (dipole) as a function of k^2. (**b**) The respective vectorwise relative errors.

4. Discussion and Conclusions

We have studied straightforward implementations of the textbook expressions for the B-field of a current loop (5), and have found that these commonly used forms are numerically troublesome at large distances from the loop and close to the axis. The computationally fast dipole approximation (8) appears to be an unsatisfactory option, as it converges very slowly and solves the problem only at large distances. Instead, a well-chosen Taylor series approximation (9) with a simple cutoff criterion can mitigate the instability globally. It is also possible to cover all of $\mathbb{R}^3$ with two overlapping Taylor series (9) and (10). However, to achieve good precision, the series must include a high order, which makes them computationally slow. A recently proposed binomial expansion (11) offers simple expressions and excellent computation performance. The only downside is the limited accuracy (more than five significant figures) in the vicinity of the loop. The real advancement of this paper is a novel, exact, numerically stable representation in terms of modified Bulirsch *cel* functions (17). Implementations based thereon offer machine precision everywhere and much faster computation times than the classical solution.

For achieving numerical stability, series approximations are always an interesting option. However, there are several arguments that speak for the proposed *cel* approach: The approximation error of the series is a result of truncation and cannot improve unless more terms are included, while the machine precision limit of the proposed *cel* algorithm depends only on the chosen floating point arithmetic. This means that a 128 bit implementation will give twice as many correct significant figures as a 64 bit implementation when using *cel*, while the series will not improve. Of course, this comes at the cost of more automatic iterations. Secondly, most approximations have a limited region of validity which requires cut-off criteria and treating multiple cases. This is problematic because in addition to programming complexity, case splitting hinders code vectorization, which can be a serious performance issue. Moreover, a *cel*-based implementation can further reduce computation time by combining the time consuming iterative parts of both field components in a single vectorized operation.

The biggest advantage of the approximation schemes is their simple form, which makes it easy to find analytic expressions for their derivatives. While it is also possible to determine derivatives of the proposed *cel* expressions, making them numerically stable seems quite laborious.

Similar stability problems of analytic forms can be found throughout the literature when systems of cylindrical symmetry are involved. Specific examples are expressions found for the axially magnetized cylinder [26], the diametrically magnetized cylinder [29], and homogeneously magnetized ring segments [30]. It is very likely that the arc solutions [17] also suffer from this problem. A similar treatment should be attempted for the extremely useful expressions in these works.

Finally, we are happy to announce that the proposed implementation of the magnetic field of a current loop, based on the cel^* and cel^{**} algorithms, was adopted into the open-access Python package Magpylib version 4.2.0 and is already available to the general public.

Author Contributions: Conceptualization, methodology, validation and investigation were performed by M.O. and P.L.; original draft preparation, visualization, administration and funding were conducted by M.O.; writing, review and editing, formal analysis were done by P.L. and F.S. All authors have read and agreed to the published version of the manuscript.

Funding: This project has been supported by the COMET K1 centre ASSIC Austrian Smart Systems Integration 100 Research Center.

Institutional Review Board Statement: Not applicable.

Informed Consent Statement: Not applicable.

Data Availability Statement: Not applicable.

Acknowledgments: This project has been supported by the COMET K1 centre ASSIC Austrian Smart Systems Integration 100 Research Center. The COMET—Competence Centers for Excellent Technologies—Program is supported by BMVIT, 101 BMDW and the federal provinces of Carinthia and Styria. Feedback and discussions with Christoph Ortner on numerical analysis were highly appreciated. Thanks to Github users @fwessel and @Blue-Spartan for pointing out numerical stability issues in the Magpylib computations. Special thanks to Alexandre Boisellet for his support of the Magpylib project.

Conflicts of Interest: The funders had no role in the design of the study; in the collection, analyses, or interpretation of data; in the writing of the manuscript; or in the decision to publish the results.

Appendix A. Algorithms

All algorithms in this section are written in Python 3.9. They are presented in a scalar form for readability.

Appendix A.1. Straightforward Implementation

The following code shows a naive implementation of the classical expressions (5) of the *B*-field of a current loop in cylindrical coordinates.

```
import math as m
from scipy.special import ellipe, ellipk

def B(rho, z):
    """
    B-field of current loop in cylindrical coordinates in units of (mT).
    """
    z2 = z**2
    x0 = (z2 + (rho+1)**2)

    k2 = 4*rho/x0
    q2 = (z2 + (rho-1)**2) / x0

    k = m.sqrt(k2)
    E = ellipe(k2)
```

```
    K = ellipk(k2)

    x1 = k/m.sqrt(rho)/q2/20
    x2 = (1+q2)*E - 2*q2*K

    Br = x1*z/rho * x2
    Bz = - x1*(x2 - k2/rho*E)

    return (Br, 0, Bz)
```

Appendix A.2. Taylor Series Implementations

The following code is an implementation of the Taylor series expansion (9a) with orders up to k^{61}.

```
import math as m

def Br_taylor_k(rho, z):
    """
    Taylor series of the radial component of the B-field of a current loop for
        small k up to order k**61.
    Inputs: Radial and axial positions , rho = rho/rho0 and z = z/rho_0.
    Output: B-field in units of millitesla
    """
    z2 = z**2
    x0 = (z2 + (rho+1)**2)
    k2 = 4*rho/x0
    q2 = (z2+(rho-1)**2)/x0
    k = m.sqrt(k2)

    P = (
        4.348319233471567e-05, # k**61
        4.660399561232684e-05,
        5.007331535430794e-05,
        5.394505893886228e-05,
        5.828397724486921e-05,
        6.316839218549437e-05, # k**51
        6.869375982346906e-05,
        7.497737506313522e-05,
        8.216465605330530e-05,
        9.043764518687763e-05,
        1.000266678574197e-04, # k**41
        1.112265650306444e-04,
        1.244196727442359e-04,
        1.401089569905765e-04,
        1.589667810126675e-04,
        1.819083471102810e-04, # k**31
        2.102052011052137e-04,
        2.456659045960062e-04,
        2.909335350495146e-04,
        3.499952301347544e-04,
        4.290963192983367e-04, # k**21
        5.384738124528147e-04,
        6.958738499390220e-04,
        9.343201341838618e-04,
        1.321260795815562e-03,
        2.013349784099904e-03, # k**11
        3.451456772742693e-03,
        7.363107781851078e-03,
        2.945243112740431e-02, # k**5
    )

    result = P[0]
    for p in P[1:]:
        result *= k2
        result += p
    result *= k2*k2*k*z/rho**(3/2)/q2

```

```
    return result
```

The following code is an implementation of the Taylor series expansion (10a) with orders up to q^{60}.

```
import math as m

def Br_taylor_q(rho, z):
    """
    Taylor series of the radial component of the B-field of a current loop for
        small q up to order q**60.
    Inputs: Radial and axial positions , rho = rho/rho0 and z = z/rho_0.
    Output: B-field in units of millitesla
    """
    z2 = z**2
    x0 = (z2 + (rho+1)**2)
    k2 = 4*rho/x0
    q2 = (z2+(rho-1)**2)/x0
    k = m.sqrt(k2)
    log_q2 = m.log(q2)

    P = (
        2.140383169962052e-08 - 2.157384139784586e-07 * log_q2, # q**60
        2.448605439302092e-08 - 2.386401724650211e-07 * log_q2,
        2.814368061953632e-08 - 2.649020096262435e-07 * log_q2,
        3.251058034393473e-08 - 2.951635428013855e-07 * log_q2,
        3.775841577054470e-08 - 3.302162609809064e-07 * log_q2,
        4.410934326064394e-08 - 3.710473232630592e-07 * log_q2, # q**50
        5.185372706900618e-08 - 4.188984109787034e-07 * log_q2,
        6.137516212651031e-08 - 4.753457145857628e-07 * log_q2,
        7.318631654988898e-08 - 5.424099936978626e-07 * log_q2,
        8.798105223936001e-08 - 6.227100551585218e-07 * log_q2,
        1.067114704413582e-07 - 7.196799136541265e-07 * log_q2, # q**40
        1.307038624714910e-07 - 8.378809805536608e-07 * log_q2,
        1.618366787740300e-07 - 9.834587570977332e-07 * log_q2,
        2.028196779163440e-07 - 1.164823878536536e-06 * log_q2,
        2.576424314628448e-07 - 1.393689567281251e-06 * log_q2,
        3.323144988153871e-07 - 1.686689932371525e-06 * log_q2, # q**30
        4.361242251684836e-07 - 2.067972330876966e-06 * log_q2,
        5.838535308384648e-07 - 2.573476678424669e-06 * log_q2,
        7.998285803570591e-07 - 3.258245255466363e-06 * log_q2,
        1.125665818448870e-06 - 4.209410019484494e-06 * log_q2,
        1.635873678758604e-06 - 5.570347093453165e-06 * log_q2, # q**20
        2.471262120850365e-06 - 7.588088919874281e-06 * log_q2,
        3.915944146725158e-06 - 1.071259612217546e-05 * log_q2,
        6.591535804299384e-06 - 1.582168042659760e-05 * log_q2,
        1.200571004791862e-05 - 2.478361129760742e-05 * log_q2,
        2.434841408803401e-05 - 4.205703735351563e-05 * log_q2, # q**10
        5.769486681760943e-05 - 8.010864257812500e-05 * log_q2,
        1.750345560741787e-04 - 1.831054687500000e-04 * log_q2,
        8.433137044373718e-04 - 5.859375000000000e-04 * log_q2,
        1.534025963549897e-02 - 4.687500000000000e-03 * log_q2,
       -6.647207708399180e-02 + 3.750000000000000e-02 * log_q2, # q**0
        5.000000000000000e-02,
    )

    result = P[0]
    for p in P[1:]:
        result *= q2
        result += p
    result *= k*z/rho**(3/2)/q2

    return result
```

The following code is an implementation of the Taylor series expansion (9b) with orders up to k^{61}.

```
import math as m

def Bz_taylor_k(rho, z):
```

```
    """
    Taylor series of the axial component of the B-field of a current loop for
        small k up to order k**61.
    Inputs: Radial and axial positions, rho = rho/rho0 and z = z/rho_0.
    Output: B-field in units of millitesla
    """
    z2 = z**2
    x0 = (z2 + (rho+1)**2)
    k2 = 4*rho/x0
    q2 = (z2+(rho-1)**2)/x0
    k = m.sqrt(k2)

    P = (
        4.348319233471567e-05 * rho + 1.499420425335023e-05, # k**61
        4.660399561232684e-05 * rho + 1.608947467568427e-05,
        5.007331535430794e-05 * rho + 1.730929419655089e-05,
        5.394505893886228e-05 * rho + 1.867328963268310e-05,
        5.828397724486921e-05 * rho + 2.020511211155466e-05,
        6.316839218549437e-05 * rho + 2.193346950885221e-05, # k**51
        6.869375982346906e-05 * rho + 2.389348167772837e-05,
        7.497737506313522e-05 * rho + 2.612847918866833e-05,
        8.216465605330530e-05 * rho + 2.869241957417010e-05,
        9.043764518687763e-05 * rho + 3.165317581540717e-05,
        1.000266678574197e-04 * rho + 3.509707644119991e-05, # k**41
        1.112265650306444e-04 * rho + 3.913527288115264e-05,
        1.244196727442359e-04 * rho + 4.391282567443621e-05,
        1.401089569905765e-04 * rho + 4.962192226749586e-05,
        1.589667810126675e-04 * rho + 5.652152213783733e-05,
        1.819083471102810e-04 * rho + 6.496726682510037e-05, # k**31
        2.102052011052137e-04 * rho + 7.545827731982029e-05,
        2.456659045960062e-04 * rho + 8.871268777078002e-05,
        2.909335350495146e-04 * rho + 1.057940127452780e-04,
        3.499952301347544e-04 * rho + 1.283315843827433e-04,
        4.290963192983367e-04 * rho + 1.589245627030877e-04, # k**21
        5.384738124528147e-04 * rho + 2.019276796698055e-04,
        6.958738499390220e-04 * rho + 2.650947999767703e-04,
        9.343201341838618e-04 * rho + 3.633467188492796e-04,
        1.321260795815562e-03 * rho + 5.285043183262248e-04,
        2.013349784099904e-03 * rho + 8.388957433749600e-04, # k**11
        3.451456772742693e-03 * rho + 1.533980787885641e-03,
        7.363107781851078e-03 * rho + 3.681553890925539e-03,
        2.945243112740431e-02 * rho + 1.963495408493621e-02,
                                    - 7.853981633974483e-02, # k**3
    )

    result = P[0]
    for p in P[1:]:
        result *= k2
        result += p
    result *= -k/rho**(3/2)/q2*k2

    return result
```

The following code is an implementation of the Taylor series expansion (10b) with orders up to q^{60}.

```
import math as m

def Bz_taylor_q(rho, z):
    """
    Taylor series of the axial component of the B-field of a current loop for
        small q up to order q**60.
    Inputs: Radial and axial positions, rho = rho/rho0 and z = z/rho_0.
    Output: B-field in units of millitesla
    """
    z2 = z**2
    x0 = (z2 + (rho+1)**2)
    k2 = 4*rho/x0
    q2 = (z2+(rho-1)**2)/x0
```

```
    k = m.sqrt(k2)
    log_q2 = m.log(q2)

    P = (
        2.140383169962052e-08 * rho + 5.756366599134164e-07, # q**60
        2.448605439302092e-08 * rho + 6.367692247077877e-07,
        2.814368061953632e-08 * rho + 7.068759571675434e-07,
        3.251058034393473e-08 * rho + 7.876663620944477e-07,
        3.775841577054470e-08 * rho + 8.812562039945226e-07,
        4.410934326064394e-08 * rho + 9.902847920909337e-07, # q**50
        5.185372706900618e-08 * rho + 1.118072627259595e-06,
        6.137516212651031e-08 * rho + 1.268835739807469e-06,
        7.318631654988898e-08 * rho + 1.447980732716223e-06,
        8.798105223936001e-08 * rho + 1.662516406570504e-06,
        1.067114704413582e-07 * rho + 1.921636483711170e-06, # q**40
        1.307038624714910e-07 * rho + 2.237557829363388e-06,
        1.618366787740300e-07 * rho + 2.626747507371085e-06,
        2.028196779163440e-07 * rho + 3.111754087360531e-06,
        2.576424314628448e-07 * rho + 3.723999925319969e-06,
        3.323144988153871e-07 * rho + 4.508141566084171e-06, # q**30
        4.361242251684836e-07 * rho + 5.529063837031901e-06,
        5.838535308384648e-07 * rho + 6.883443473579986e-06,
        7.998285803570591e-07 * rho + 8.719539805210148e-06,
        1.125665818448870e-06 * rho + 1.127244058408705e-05,
        1.635873678758604e-06 * rho + 1.492981081843003e-05, # q**20
        2.471262120850365e-06 * rho + 2.036144759732213e-05,
        3.915944146725158e-06 * rho + 2.879192411774280e-05,
        6.591535804299384e-06 * rho + 4.262260553943059e-05,
        1.200571004791862e-05 * rho + 6.700276866917860e-05,
        2.434841408803401e-05 * rho + 1.143628488115506e-04, # q**10
        5.769486681760943e-05 * rho + 2.201260551956201e-04,
        1.750345560741787e-04 * rho + 5.143424513214410e-04,
        8.433137044373718e-04 * rho + 1.748691113312115e-03,
        1.534025963549897e-02 * rho + 1.931709939249829e-02,
       -6.647207708399180e-02 * rho + 2.784264097200273e-02, # q**0
        5.000000000000000e-02 * rho - 5.000000000000000e-02, # q**-2
    )
    Q = (
        2.157384139784586e-07 * rho + 8.701449363797829e-06, # q**60
        2.386401724650211e-07 * rho + 9.306966726135822e-06,
        2.649020096262435e-07 * rho + 9.977975695921838e-06,
        2.951635428013855e-07 * rho + 1.072427538845034e-05,
        3.302162609809064e-07 * rho + 1.155756913433172e-05,
        3.710473232630592e-07 * rho + 1.249192654985633e-05, # q**50
        4.188984109787034e-07 * rho + 1.354438195497808e-05,
        4.753457145857628e-07 * rho + 1.473571715215865e-05,
        5.424099936978626e-07 * rho + 1.609149647970326e-05,
        6.227100551585218e-07 * rho + 1.764345156282479e-05,
        7.196799136541265e-07 * rho + 1.943135766866142e-05, # q**40
        8.378809805536608e-07 * rho + 2.150561183421063e-05,
        9.834587570977332e-07 * rho + 2.393082975604484e-05,
        1.164823878536536e-06 * rho + 2.679094920634033e-05,
        1.393689567281251e-06 * rho + 3.019660729109378e-05,
        1.686689932371525e-06 * rho + 3.429602862488768e-05, # q**30
        2.067972330876966e-06 * rho + 3.929147428666235e-05,
        2.573476678424669e-06 * rho + 4.546475465216915e-05,
        3.258245255466363e-06 * rho + 5.321800583928393e-05,
        4.209410019484494e-06 * rho + 6.314115029226741e-05,
        5.570347093453165e-06 * rho + 7.612807694385992e-05, # q**20
        7.588088919874281e-06 * rho + 9.358643001178280e-05,
        1.071259612217546e-05 * rho + 1.178385573439300e-04,
        1.582168042659760e-05 * rho + 1.529429107904434e-04,
        2.478361129760742e-05 * rho + 2.065300941467285e-04,
        4.205703735351563e-05 * rho + 2.943992614746094e-04, # q**10
        8.010864257812500e-05 * rho + 4.539489746093750e-04,
        1.831054687500000e-04 * rho + 7.934570312500000e-04,
        5.859375000000000e-04 * rho + 1.757812500000000e-03,
        4.687500000000000e-03 * rho + 7.812500000000000e-03,
       -3.750000000000000e-02 * rho - 1.250000000000000e-02, # q**0
```

```
            0.,
        )

    result = P[0] - Q[0]*log_q2
    for p,q in zip(P[1:], Q[1:]):
        result *= q2
        result += p - q*log_q2
    result *= -k/rho**(3/2)/q2

    return result
```

In most cases, the 30 provided terms will not be necessary. A relation between the number of terms included and precision of the Taylor series is shown in Table A1. The table gives values of k below (k-series) or above (q-series) which a series has a precision below the indicated number of significant figures when the indicated order is included in the sum. For example, an implementation of Br_taylor_k that includes up to order k^{21} has at least 8 digits of precision when $k \leq 0.46$. An implementation of Br_taylor_q that includes up to order q^{40} has at least 10 digits of precision when $k \geq 0.69$.

Table A1. Precision of various Taylor implementations in relation to the number of terms included in the respective sums.

	Br_taylor_k ($k \leq$ value)						**Br_taylor_q** ($k \geq$ value)				
sig.figs	k-order included in series					sig.figs	q-order included in series				
	21	31	41	51	61		20	30	40	50	60
8	0.46	0.62	0.71	0.77	0.81	8	0.77	0.67	0.60	0.55	0.50
10	0.35	0.53	0.63	0.70	0.75	10	0.85	0.76	0.69	0.63	0.58
12	0.27	0.45	0.56	0.64	0.69	12	0.90	0.82	0.75	0.70	0.65
	Bz_taylor_k ($k \leq$ value)						**Bz_taylor_q** ($k \geq$ value)				
sig.figs	k-order included in series					sig.figs	q-order included in series				
	21	31	41	51	61		20	30	40	50	60
8	0.43	0.59	0.68	0.74	0.78	8	0.86	0.76	0.70	0.64	0.59
10	0.33	0.50	0.61	0.68	0.73	10	0.91	0.83	0.76	0.71	0.66
12	0.26	0.43	0.54	0.62	0.68	12	0.94	0.87	0.81	0.76	0.72

Appendix A.3. Original Implementation of Bulirsch's cel Algorithm

The original implementation of Bulirsch's *cel* algorithm [25] is reproduced here. This implementation is also found in the first edition of the popular Numerical Recipes [31] and is commented on in [26]. Advantages of the algorithm over representations in terms of Carlson's functions are discussed in [32]. A slightly faster implementation by Fukushima is found in [27].

```
import math as m

def cel(kc, p, c, s):
    """
    Bulirsch's cel algorithm taken from Derby2010
    """
    if kc == 0:
        raise RuntimeError("FAIL")
    errtol = 0.000001
    k = abs(kc)
    pp = p
    cc = c
    ss = s
    em = 1.0
    if p > 0:
        pp = m.sqrt(p)
        ss = s / pp
```

```
    else:
        f = kc * kc
        q = 1.0 - f
        g = 1.0 - pp
        f = f - pp
        q = q * (ss - c * pp)
        pp = m.sqrt(f / g)
        cc = (c - ss) / g
        ss = -q / (g * g * pp) + cc * pp
    f = cc
    cc = cc + ss / pp
    g = k / pp
    ss = 2 * (ss + f * g)
    pp = g + pp
    g = em
    em = k + em
    kk = k
    while abs(g - k) > g * errtol:
        k = 2 * m.sqrt(kk)
        kk = k * em
        f = cc
        cc = cc + ss / pp
        g = kk / pp
        ss = 2 * (ss + f * g)
        pp = g + pp
        g = em
        em = k + em
    return (m.pi / 2) * (ss + cc * em) / (em * (em + pp))
```

*Appendix A.4. Implementation of cel**

Here, we show a modification of Bulirsch's *cel* algorithm for computing the function *cel**, described in Section 2.7. Input arguments are k^2 and q^2. A vectorized version of this algorithm is provided by the open-source library Magpylib [11].

```
import math as m

def celx(k2, q2):
    """
    Modified Bulirsch cel algorithm for computing
    xi0 = (1+q2)*E(k2)-2*q2*K(k2)
        = cel(q, 1, k2, -k2*q2)
    with 0 < k2 < 1, and q2 = 1-k2
    """
    qc = m.sqrt(q2)
    p  = 1 + qc
    g  = 1
    cc = k2 * k2
    ss = 2 * cc * (qc / (qc + 1) )
    em = p
    kk = qc

    while m.fabs(g - qc) >= g * 1e-8:
        qc = 2 * m.sqrt(kk)
        kk = qc * em
        f  = cc
        cc = cc + ss / p
        g  = kk / p
        ss = 2 * (ss + f * g)
        p  = p + g
        g  = em
        em = em + qc
    return 1.5707963267948966 * (ss + cc * em) / (em * (em + p))
```

Appendix A.5. Implementation of cel**

Here, we show a modification of Bulirsch's *cel* algorithm for computing the function *cel***, described in Section 2.7. Input arguments are $\tilde{\rho}$, k^2 and q^2. A vectorized version of this algorithm is provided by the open-source library Magpylib [11].

```
import math as m

def celxx(rho, k2, q2):
    """
    Modified Bulirsch cel algorithm for computing
    xi1 = (1+q2-k2/rho)*E(k2) - 2*q2*K(k2)
        = cel(q, 1, k2*(1-1/rho), -k2*q2*(1+1/rho))
    with 0 < k2 < 1, and q2 = 1-k2
    """
    qc = m.sqrt(q2)
    p = 1 + qc
    g = 1
    cc = k2*(k2-(q2+1)/rho)
    ss = 2 * k2 * qc * (k2/(1+qc) - (qc+1)/rho)
    em = p
    kk = qc

    while m.fabs(g - qc) >= g * 1e-8:
        qc = 2 * m.sqrt(kk)
        kk = qc * em
        f  = cc
        cc = cc + ss / p
        g  = kk / p
        ss = 2 * (ss + f * g)
        p  = p + g
        g  = em
        em = em + qc
    return 1.5707963267948966 * (ss + cc * em) / (em * (em + p))
```

References

1. Madenci, E.; Guven, I. *The Finite Element Method and Applications in Engineering Using ANSYS®*; Springer: Berlin/Heidelberg, Germany, 2015.
2. Pryor, R.W. *Multiphysics Modeling Using COMSOL®: A First Principles Approach*; Jones & Bartlett Publishers: Burlington, MA, USA, 2009. 2009.
3. Alnæs, M.; Blechta, J.; Hake, J.; Johansson, A.; Kehlet, B.; Logg, A.; Richardson, C.; Ring, J.; Rognes, M.E.; Wells, G.N. The FEniCS project version 1.5. *Arch. Numer. Softw.* **2015**, *3*, 9–23.
4. Schöberl, J. *C++ 11 Implementation of Finite Elements in NGSolve;* Institute for Analysis and Scientific Computing, Vienna University of Technology: Vienna, Austria 2014; Volume 30.
5. Smythe, W.B. *Static and Dynamic Electricity*; Hemisphere Publishing: New York, NY, USA, 1988.
6. Moshier, S.L.B. *Methods and Programs for Mathematical Functions*; Ellis Horwood Ltd Publisher: Chichester, UK, 1989.
7. ALGLIB. Available online: https://www.alglib.net/download.php (accessed on 9 June 2022).
8. Wolfram, S. *The Mathematica Book*; Wolfram Research, Inc.: Champaign, IL, USA 2003; Volume 1.
9. MATLAB. *Version 7.10.0 (R2010a)*; The MathWorks Inc.: Natick, MA, USA, 2010.
10. Virtanen, P.; Gommers, R.; Oliphant, T.E.; Haberland, M.; Reddy, T.; Cournapeau, D.; Burovski, E.; Peterson, P.; Weckesser, W.; Bright, J.; et al. SciPy 1.0: Fundamental algorithms for scientific computing in Python. *Nat. Methods* **2020**, *17*, 261–272.
11. Ortner, M.; Bandeira, L.G.C. Magpylib: A free Python package for magnetic field computation. *SoftwareX* **2020**, *11*, 100466.
12. Higham, N.J. *Accuracy and Stability of Numerical Algorithms*; SIAM: Philadelphia, PA, USA, 2002.
13. Jackson, J.D. Classical electrodynamics. *Am. J. Phys.* **1999**, *67*, 841.
14. Ortner, M.; Filipitsch, B. Feedback of Eddy Currents in Layered Materials for Magnetic Speed Sensing. *IEEE Trans. Magn.* **2017**, *53*, 1–11.
15. Simpson, J.C.; Lane, J.E.; Immer, C.D.; Youngquist, R.C. *Simple Analytic Expressions for the Magnetic Field of a Circular Current Loop*; Technical Report; NASA, Kennedy Space Center: Merritt Island, FL, USA, 2001.
16. Behtouei, M.; Faillace, L.; Spataro, B.; Variola, A.; Migliorati, M. A novel exact analytical expression for the magnetic field of a solenoid. *Waves Random Complex Media* **2020**, *32*, 1977–1991.
17. González, M.A.; Cárdenas, D.E. Analytical Expressions for the Magnetic Field Generated by a Circular Arc Filament Carrying a Direct Current. *IEEE Access* **2020**, *9*, 7483–7495.
18. Prantner, M.; Parspour, N. Analytic multi Taylor approximation (MTA) for the magnetic field of a filamentary circular current loop. *J. Magn. Magn. Mater.* **2021**, *517*, 167365.

19. Chapman, G.H.; Carleton, D.E.; Sahota, D.G. Current Loop Off Axis Field Approximations with Excellent Accuracy and Low Computational Cost. *IEEE Trans. Magn.* **2022**, *58*, 1–6.
20. Seleznyova, K.; Strugatsky, M.; Kliava, J. Modelling the magnetic dipole. *Eur. J. Phys.* **2016**, *37*, 025203.
21. Schill, R.A. General relation for the vector magnetic field of a circular current loop: A closer look. *IEEE Trans. Magn.* **2003**, *39*, 961–967.
22. Urzhumov, Y.; Smith, D.R. Metamaterial-enhanced coupling between magnetic dipoles for efficient wireless power transfer. *Phys. Rev. B* **2011**, *83*, 205114.
23. Rong, Z.; Wei, Y.; Klinger, L.; Yamauchi, M.; Xu, W.; Kong, D.; Cui, J.; Shen, C.; Yang, Y.; Zhu, R.; et al. A New Technique to Diagnose the Geomagnetic Field Based on a Single Circular Current Loop Model. *J. Geophys. Res. Solid Earth* **2021**, *126*, e2021JB022778.
24. Alldredge, L.R. Circular current loops, magnetic dipoles and spherical harmonic analyses. *J. Geomagn. Geoelectr.* **1980**, *32*, 357–364.
25. Bulirsch, R. Numerical calculation of elliptic integrals and elliptic functions. III. *Numer. Math.* **1969**, *13*, 305–315.
26. Derby, N.; Olbert, S. Cylindrical magnets and ideal solenoids. *Am. J. Phys.* **2010**, *78*, 229–235.
27. Fukushima, T.; Kopeikin, S. Elliptic functions and elliptic integrals for celestial mechanics and dynamical astronomy. *Front. Relativ. Celest. Mech.* **2014**, *2*, 189–228.
28. Fukushima, T.; Ishizaki, H. Numerical computation of incomplete elliptic integrals of a general form. *Celest. Mech. Dyn. Astron.* **1994**, *59*, 237–251.
29. Caciagli, A.; Baars, R.J.; Philipse, A.P.; Kuipers, B.W. Exact expression for the magnetic field of a finite cylinder with arbitrary uniform magnetization. *J. Magn. Magn. Mater.* **2018**, *456*, 423–432.
30. Slanovc, F.; Ortner, M.; Moridi, M.; Abert, C.; Suess, D. Full analytical solution for the magnetic field of uniformly magnetized cylinder tiles. *J. Magn. Magn. Mater.* **2022**, *559*, 169482.
31. Press, W.H.; Teukolsky, S.A.; Vetterling, W.T.; Flannery, B.P. *Numerical Recipes 3rd Edition: The Art of Scientific Computing*; Cambridge University Press: Cambridge, UK, 2007.
32. Reinsch, K.D.; Raab, W. Elliptic Integrals of the First and Second Kind—Comparison of Bulirsch's and Carlson's Algorithms for Numerical Calculation. In *Special Functions*; World Scientific: Singapore 2000; pp. 293–308.

magnetism

MDPI

Article

Practical Study of Mixed-Core High Frequency Power Transformer

Arun Kumar Paul

Research and Development, M/S. Electronics Devices Worldwide Pvt. Ltd., 22 Mistry Industrial Estate, Cross Road A, Andheri East, Mumbai 400093, India; arunp26@iitbombay.org

Abstract: The design of medium- to high-frequency power electronics transformer aims not only to minimize the power loss in the windings and the core, but its heat removal features should also allow optimal use of both core and copper. The heat removal feature (e.g., thermal conduction) of a transformer is complex because there exist multiple loss centers. The bulk of total power loss is concentrated around a small segment of the core assembly where windings are overlaid. The primary winding is most constrained thermally. For superior use of core and copper, the temperature rise in different segments of the transformer should be well below their respective safe operating limits. In practice, cores of same soft-magnetic materials are traditionally used. To achieve superior temperature profile and for better long-term performance, this article proposes to use the mixed-core configuration. The new core(s) would replace the parent ones from the segment where windings are laid. The characteristic features of new cores would share increased burden of heat removal from the transformer. To obtain the qualitative insight of magnetic and thermal performance, the proposed mixed-core transformer would be thoroughly validated practically in two different high-power applications. In the first case, the core is always energized to its rated value, and in the second one, windings are always energized at respective rated current capacity.

Keywords: hot spot temperature; mixed-core magnetic circuit; power electronics systems; power electronic transformer (PET); soft magnetic materials

1. Introduction

Power electronics converters [1] are used to ensure efficient handling of electrical energy where, as shown in Figure 1, power electronic transformer (PET) is an integral part [2–7]. Though it is a passive component, its role in power converters is immense. Apart from mandatory safety isolation, a PET could be used either for voltage [8] or current multiplication [4] to match the load characteristics of applications. Depending upon the topology of power converter used, additionally, it is often used to perform certain assisting roles in different soft-switched inverter operations [9–11]. The PET is used for instantaneous power transfer, and it should be efficient and compact. Due to the availability of a wide range of soft magnetic materials [12,13] in different geometries as well as of copper conductors (litz wires and copper foils) [14], the process of design optimization of a PET is now elaborate. For optimal design of a PET, the following aspects need to be looked into:

1. The power loss characteristics and its distribution in core and copper [15–21].
2. The distribution of steady state temperature in different parts of the core and copper windings [3,21].
3. The influence of soft magnetic materials of suitable geometry [13] as well as that of copper conductors.
4. The leakage inductance [11,22].
5. The prospect of static and dynamic DC bias in the PET [8,17], etc.

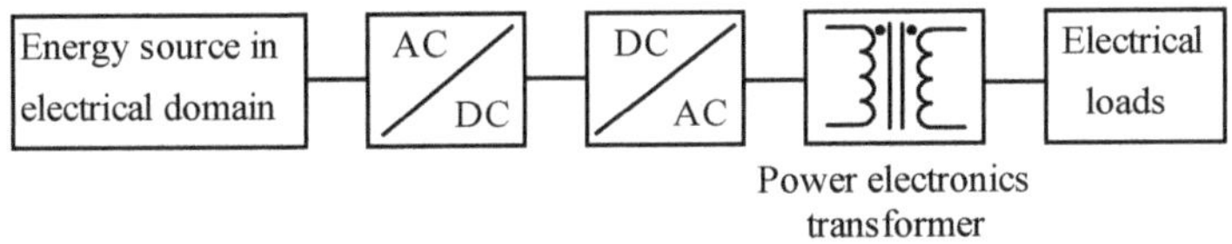

Figure 1. Arrangement of a basic power electronics system.

One typical 2-winding PET is shown in Figure 2. Its design optimization involves optimal use of core and copper; it is possible if both the core and copper losses are minimized. The copper loss P_{cu} depends on the current density of conductors, the impact of proximity effects on the resistance values of primary and secondary conductors, winding configuration and the construction of conductors [14,23]. On the other hand, the core loss P_{core} depends on the properties of the core material, peak operating flux density B_m, the excitation frequency f_s, the waveform pattern and the core temperature [18,19,24,25]. Thirdly, the design also involves devising a thermal circuit so that the temperature rise in core, copper and insulation are not only within the respective safe operating limit, but there should also be increased uniformity of maximum temperature rise in different parts of windings as well as in the core. Ensuring near-uniform temperature rise is complex because the distribution of power loss in the core is not uniform and so is the case for windings. Moreover, the thermal behavior of each circuit is also different. Ideally, for the design of the thermal circuit for heat removal, the average flux density per cycle is considered to be zero where the core loss is decided by the values of B_m and f_s; it is true when the DC bias in core is absent [17]. The presence of DC bias could adversely affect the performance of the PET in several ways. Primarily, the core loss increases significantly under DC bias; it could work as a hindrance to draw any comparative statement on performance among different PETs. Secondly, depending upon the DC bias capacity of the magnetic circuit, there could be core saturation that affects the performance of the power controller. The DC bias capacity of the magnetic circuit is poor for zero-gap magnetic circuit using high permeability materials (e.g., toroidal core using nanocrystalline materials). The DC bias could be static [26] or dynamic [27,28].

Figure 2. Typical power transformer using EE or UU cores.

Except in a few applications [8] where certain leakage inductance is desired, the design of PET necessitates that the two windings are overlaid on a small segment of the core volume [29]. Therefore, the bulk of its total power loss is concentrated around a small core volume (e.g., the central limb of Figure 2). Removal of such concentrated heat loss is a complex task. It would decide the operating limits of the core and the windings [21]. Large, concentrated power loss along with the constrained heat removal features of core and windings would result nonuniform temperature rise [3]. The respective value of maximum temperature rise would decide the capacity of core and windings. It forces the core to operate at reduced flux density because the maximum permissible operating temperature of core is less than that of copper or its insulation. It would affect the power density of the

PET. To get the desired flux level, several cores are integrated. Until recently, for medium- to high-frequency PETs, ferrites have been prominently used to build the magnetic circuit. As shown in Table 1, for ferrites, the heat removal by thermal conduction is not attractive where the prospect of creation of hot spot is comparatively large [20]. The safe operating temperature range of ferrites is also not large and several of its parameters are sensitive to temperature [13]. Therefore, cores are usually integrated to achieve multiple goals, e.g., to meet the desired flux level, to have improved thermal conduction features to distribute the core loss over larger core surface area, to reduce the number of turns and layers in the windings, etc.

Table 1. Comparison on basic parameters of a few soft magnetic materials for transformer.

Core Material	Si-Steel	Ferrite	Amorphous	Nanocrystalline
Thermal conductivity K, °C/mK	18.6	≤5	10	10
Loss density (W/kg) at 0.1 T, 20 kHz	20.0	1.9	4.7	0.9
Loss density (kW/m^3) at 0.1 T, 20 kHz	150	9.1	34	6.6
Relative permeability μ_r	900	2200	15,000	15,000
Maximum operating temperature, °C	150	100	150	120–150
Saturation flux density B_{sat} at 25 °C	1.8	4.9	1.56	1.23
Curie Temperature, °C	770	230	415	570

As detailed in [3], the temperature rise in various segments of core and copper is different; the safe operating temperature limit of each component of a PET could as well be different. For any power electronic component, for example, the differential value between its limiting temperature and maximum (hot spot) operating temperature would decide its utility as well as the service life [30]. The prospect of creation of hot spot temperature in copper is high on winding close to the core. The temperature of core is also maximum there [3]. The value of thermal conductivity K of core could play an important role in controlling the temperature distribution because, as shown in Figure 2, a major part of core assembly remains exposed to the surrounding environment. The bulk of power loss in a PET is concentrated around a small part of the core where primary and secondary windings are overlaid. Traditionally, to share the magnetic burden (flux density, core loss) equally, several cores of a similar type have been used to integrate the flux in the magnetic circuit and reduce the core loss density. Laterally, it would reduce the thermal resistance of the magnetic circuit. Often, for parametric matching dynamically, the. same batch code of cores has been strongly recommended. Now, wide-range soft magnetic materials possessing different parametric values are available. To improve the utility of a PET, can different core materials [12,13,25,31,32] be integrated in the magnetic circuit? What could be their characteristic features for integration in series or parallel configuration in a magnetic circuit to reduce either the power loss and/or to improve the thermal behavior of the PET?

The inspiration for this experiment-driven research has been the work reported in [3]. Using finite element analysis (FEA) and validated by requisite practical demonstration, it could correctly estimate the temperature distribution in different segments of core and copper. With the aim to improve the performance of PET, this article proposes, with detailed practical demonstration, to integrate different soft magnetic materials into the magnetic circuit. Here, an improved thermal conductivity and/or reduced loss density of the new core placed in a zone of high power loss is aimed to aid the cooling of regions of maximum heat generation. The structure of the article is organized as follows: Section 2 details the features of PETs for two different characteristics applications—one used for voltage ratio and the other for current multiplication. It also discusses their design issues. Section 3 details the different perspectives of two types of mixed-core configuration suitable for two different applications. To take care of important issues such as the magnetic compatibility, etc., it also introduces application specific suitability factor for integrating different soft magnetic materials into the core. Finally, Section 4 details the practical validation of mixed-core, air-cooled transformers for two different application domains.

2. Power Electronics Transformer for Divergent Load Characteristics

Any new idea needs to be validated in different application domains. A PET is used to make the load compatible to the source; it could be achieved in several ways. A PET could be part of power transfer in PWM controlled full-bridge DC−DC converters (FBDC) [21,33], resonant converters [6,34] or for feeding a resonant tank circuit [4]. The nature of waveforms of current and voltage could change in applications. The PET could be used for voltage ratio where, at zero output power, the current in windings could be zero, or it could be used for current ratio where its primary voltage is decided by the load. In the first case, the magnetic circuit could remain loaded and the value of copper loss P_{cu} could be zero at no load. In the second case, irrespective of the magnitude of the delivered power, the windings always draw the set current. The value of the core loss P_{core} is negligible at zero power.

Achieving optimum power density of a PET is a major design goal [35]. It depends on the total power loss and the design of thermal circuit. The PET is said to be thermally loaded to its capacity when the rated power, say, P_{PET}, is delivered to the load. Traditionally, the area product $A_{\mathrm{W}}A_{\mathrm{c}}$ in Equation (1) is used to define the extent of optimization of a PET. At a particular frequency f_s, it suggests a large value of B_{m} in core as well as the current density J in copper windings, as given below,

$$A_{\mathrm{W}}A_{\mathrm{c}} \propto \frac{P_{\mathrm{PET}}}{B_{\mathrm{m}}J} \tag{1}$$

The expression of B_{m} for a square wave input voltage V_{in} is,

$$B_{\mathrm{m}} = \frac{V_{\mathrm{in}}}{4n_{\mathrm{p}}A_{\mathrm{c}}f_s} \tag{2}$$

A_{w} is the window area, A_{c} is the core area and n_{p} is the number of primary turns.

For effective use of a PET, its core loss P_{core} [18] and copper loss P_{cu} [23] need to be calculated accurately. For sinusoidal primary voltage, the Steinmetz equation [16] is used to calculate the value of P_{core}; its parameters are mostly mentioned in core datasheet. However, in high-frequency applications, the primary voltage is rarely sinusoidal. Using the same Steinmetz parameters, the improved generalized Steinmetz equation (iGSE) is used to calculate P_{core} for any input voltage waveform [17,19]. Using the iGSE, the expression of P_{core} with square wave excitation is,

$$P_{\mathrm{core}} = W_{\mathrm{c}}K_I 2^{(\alpha+\beta)} f_s^{\alpha} B_m^{\beta} d_{\mathrm{pwm}}^{(1-\alpha)} \tag{3}$$

where $K_I = \frac{K_S}{2^{(\beta+1)}\pi^{(\alpha-1)}\left(0.2761+\frac{1.7061}{\alpha+1.354}\right)}$.

K_S, α and β are Steinmetz parameters, d_{pwm} is duty cycle of square wave input and W_{c} is the core weight. For pure square wave input (d_{pwm}: 1.0), Equation (3) may be modified to,

$$P_{\mathrm{core}} = W_{\mathrm{c}}K_{S1}f_s^{\alpha}B_m^{\beta}, \quad \text{where,} \quad K_{S1} = K_I 2^{(\alpha+\beta)} \tag{4}$$

The expressions of $P_{\mathrm{cu}} = P_{\mathrm{pri}} + P_{\mathrm{sec}}$ in primary (P_{pri}) and secondary (P_{sec}) windings are,

$$P_{\mathrm{pri}} = F_1 r_{\mathrm{dc1}} i_{\mathrm{p}}^2 \text{ and } P_{\mathrm{sec}} = F_2 i_{\mathrm{s}}^2 r_{\mathrm{dc2}} \tag{5}$$

F_1 and F_2 are the ac resistance factors, i_p and i_s are the primary and the secondary current, respectively, and r_{dc1} and r_{dc2} are their respective dc resistance values. Both F_1 and F_2 depend on several factors such as skin and proximity effects where proper choice of copper conductors (litz wire or thin foil) and layout of windings are important [23].

The popular geometry of the magnetic circuit could be based on any of EE, UU or CC as shown in Figure 3a, or zero-gap toroidal shaped cores of Figure 3b. The dynamic profile of the input voltage V_{in} decides the value of B_m in core; it depends on the power controller and the characteristics of the connected load. Depending upon the value of B_m, the magnetic circuit could face nonlinearity as well as the magnetic saturation. Here, two application types are considered where the dynamics of V_{in} or B_m are completely different.

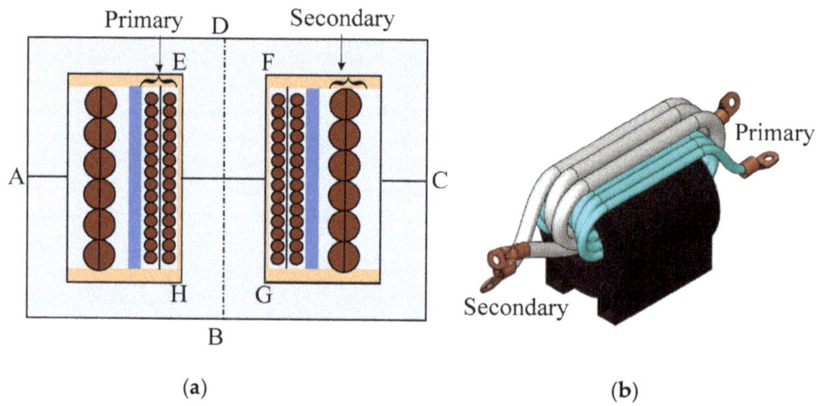

Figure 3. Popular arrangements of PET where magnetic circuit is configured in (**a**) series reluctance circuit model and (**b**) parallel reluctance circuit model.

2.1. PET for Full-Bridge DC−DC Converter (FBDC)

One typical circuit of FBDC is shown in Figure 4. Here, the magnitude of output current depends on the power drawn by the applied load at voltage V_L and its effective resistance, e.g., battery charging [11], arc welding [33], etc. Even at zero load current the cores could be fully loaded. For a nonlinear load (e.g., welding arc), the dynamic control of DC current I_a would decide the value of V_{in} or B_m through dynamic change in d_{pwm}, such as,

$$\frac{d}{dt}(I_a) = \frac{1}{L_1}(k_1 u - V_L) = \frac{1}{L_1}\left(\frac{V_{DC}}{n} d_{pwm} - V_L\right) = \frac{1}{L_1}\left(\frac{V_{in}}{n} - V_L\right) \tag{6}$$

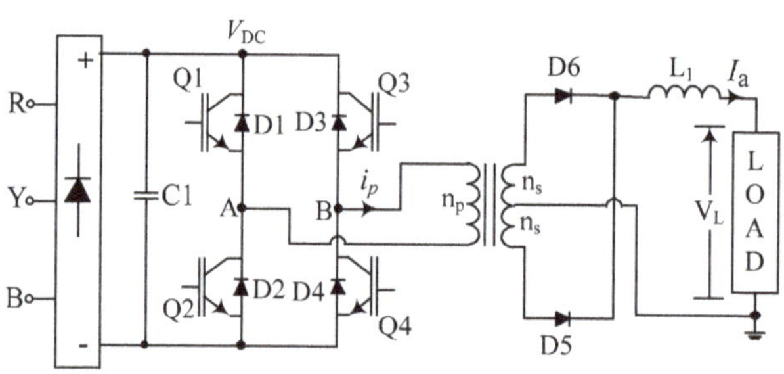

Figure 4. Typical power circuit of full-bridge DC−DC converter.

V_{DC} is the supply voltage, $n = n_p/n_s$ is the ratio of primary (n_p) to secondary (n_s) turns, k_1 is constant and the control $u = f(e)$ is used to ensure zero current error. The transient disturbance in the arc welding process is large [28].

The permissible value of J in windings would be decided by several factors such as values of F_1, F_2 and d_{pwm}. For the arrangement of secondary side rectifier of Figure 4, the primary current i_p and the current i_s in each bifilar secondary could be expressed as,

$$i_p = \frac{I_a}{n}\sqrt{d_{pwm}} \text{ and } i_s = \sqrt{\frac{1}{2}}I_a \tag{7}$$

If care is not taken, there could be error in P_{core} because the core in FBDC often faces both the static or dynamic DC bias. The static DC bias could be compensated by a simple approach [26]. The dynamic DC bias [27,28] in core would depend on how the control u (or d_{pwm}) reacts to ripple in steady state error as well as the load transients. The DC bias is more prominent when the loop gains are large where the ripple in I_a becomes transparent in control input u [28]. Under DC bias conditions, the values of Steinmetz parameters drifts.

2.2. PET for Series Resonant Induction Heating Equipment

The characteristics of the connected load to PET as well as its associated dynamics is quite different in induction heating. Here, the coil head L4 is kept energized with rated current i_L at the frequency decided by the tank circuit parameters L4 and Cr. The coil facilitates the power transfer when a metallic object is taken close to the coil. For a noncontact mode of power transfer, normally, the coil current i_L is kept large. To reduce the stress on primary side components, as shown in Figure 5, the induction heating transformer (IHT) is used to step up the inverter current. The value of P_{cu} is always at its rated value. The loading of its magnetic circuit and hence the core loss P_{core} depends on the power P_{OUT} drawn through L4; it depends on multiple parameters, such as,

$$P_{OUT} = K_c L_4 i_L^2 f_s = i_L^2 R_{eq} \tag{8}$$

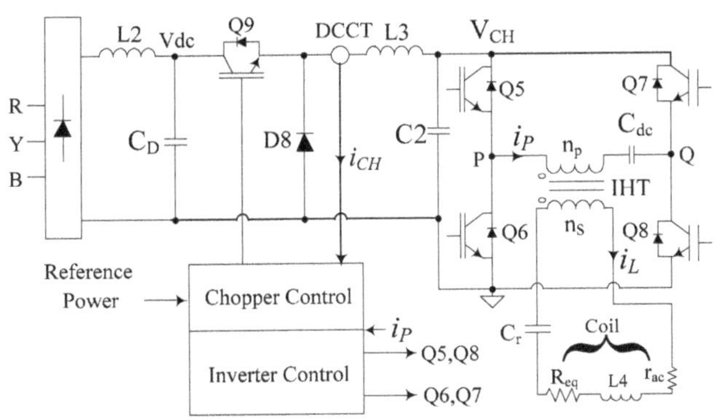

Figure 5. The power circuit for a series resonant induction heating controller.

L4 is the inductance value of coil, i_L is the coil current, f_s is frequency of i_L and the parameter K_c depends on coupling between the coil and the load. R_{eq} represents the effective load resistance reflected in the tank circuit.

The input voltage to IHT is a square wave $(d_{pwm} = 1)$, but the current in both the windings are pure sinusoidal. The DC blocking capacitor C_{dc} is connected to eliminate any static DC bias present in the core. Zero voltage switching (ZVS) condition of switches Q5–Q8 is inherently achieved because the phase-locked loop (PLL) ensures f_s at slightly higher than the resonant frequency $f_r = 1/2\pi\sqrt{L_4C_r}$. Plus, the buck converter controls the input voltage V_{in} to achieve near-zero current switching of Q5–Q8. Under ZVZCS condition, the inverter input voltage V_{CH} or the primary voltage of IHT could be approximated as,

$$V_{CH} \approx V_{in} \approx 0.787n(r_{ac} + R_{eq})|i_L| \tag{9}$$

where r_{ac} is the ac resistance of L4 and n is the turns ratio of IHT. At no load, the value of B_m is negligible because the value of $V_{in} \approx 0.787 n r_{ac}|i_L|$ is small. The change in load of IHT, i.e., the change in R_{eq} is never abrupt. Its value increases when a job is brought close to the coil (i.e., when more power is drawn through L4), and decreases gradually either near the Curie point or when the job is taken away from the coil head mechanically. They ensure that the dynamic change in B_m is also not abrupt. Moreover, the response time of the buck chopper decides the dynamics of V_{in}. Therefore, the prospect of dynamic DC bias in IHT would be small. Furthermore, the slow dynamic DC bias in the core, if any, could be effectively tackled in the PLL loop [36].

The design of IHT involves deciding on the values of n_p and n_s; selecting a core material of suitable geometry to afford optimal values of B_m and J. In ZVZCS conditions, the value of n would be maximum at n_{max} because the load behaves as resistive,

$$n = n_{max} = \frac{n_p}{n_s(\min)} = 1.27\frac{V_{CH(max)}}{(r_{ac} + R_{eq})|i_L|} \approx 1.27\frac{V_{DC}}{(r_{ac} + R_{eq})|i_L|} \tag{10}$$

At $n = n_{max}$, the current is minimum in the primary winding and the length of conductor used in the secondary is also minimized, and together they help minimize P_{cu}. Large values of B_{sat} and small values of P_c of nanocrystalline cores would allow optimal choice of n_p as well.

Secondly, for efficient transfer of power $P_{OUT} = i_L^2 R_{eq}$, tracking of f_r should be accurate. The value of R_{eq} would be more at higher values of f_r [37]. When L4 is loaded, its inductance value drifts down to, say, L_{eq}; then, the corresponding value of f_r is,

$$f_r = \frac{1}{2\pi\sqrt{(L_{lk} + L_{eq})C_r}} \approx f_s \tag{11}$$

L_{lk} is leakage inductance of IHT; its large value would be a hindrance to effective power transfer [37]. The value of L_{lk} is small for high-permeability ungapped toroidal cores.

Thirdly, the primary current i_{pri} consists of triangular wave magnetizing current with peak at I_m plus the reflected sinusoidal coil current i_L. For minimum phase error between V_{pri} and i_{pri} of IHT, the value of I_m should be small; it is expressed as,

$$I_m = \frac{l_m B_m}{\mu_0 \mu_r n_p} \tag{12}$$

where l_m is the mean core length. A large value of μ_r is needed for small value of I_m and n_p.

For IHT, the high-permeability nanocrystalline material-based ungapped toroidal cores (shown in Figure 3b) would be preferred [15,38–40], particularly because the DC bias in the core is negligible. Laterally, these cores would ensure small values of I_m and L_{lk} and also the minimum of number of turns where the value of B_m would be large.

3. Mixed-Core Transformer Configuration

It was clear in Sections 2.1 and 2.2 that the type of application or load characteristics could influence the design of PET. The nature of loading of magnetic circuit, in particular, could vary in applications, e.g., the value of P_{core} could be fixed and that of P_{cu} would be decided by the load. Along with reducing the core and the copper losses, the optimization process involves design of a thermal circuit to ensure near-uniform temperature rise in core and also in copper so that the PET is enabled to deliver more power. Due to multiple loss centers, the thermal circuit of the core and windings are coupled. For the magnetic circuit, the distribution of heat and its removal by thermal convection could be improved if the value of K as well as that of the surface area of core are increased. To have requisite flux $A_c B_m$, several cores are integrated. Often, for magnetic compatibility, cores of the same material with the same batch code are preferred. It is important to find whether such arrangement is best suitable for efficient heat removal, both from the core assembly as well as from the windings. On the other hand, can some other combinations, such as the hybrid core configuration, manage the heat loss or the thermal issues better?

3.1. Thermal Behavior of Power Electronic Transformer

As shown in Equations (3) and (5), the values of P_{core} and P_{cu}, respectively, increase exponentially with B_m and with J^2. The design optimization of a PET is complex because the layout of the windings and thermal behavior of the PET often contradict. As shown in Figure 3a,b, two windings are overlaid for better magnetic coupling and also for reduced eddy current loss in core [29]. Such arrangement needs good heat removal features because the bulk of total power loss ($P_{tot} = P_{core} + P_{cu}$) is concentrated around a small core segment where the secondary winding is laid above the primary. Removing the heat loss from the multilayered primary winding is difficult because there exists insulation on either side of each layer and also between the windings. The prospect of creation of hot spot is more in the primary [3]. Furthermore, due to the increased impact of the proximity effect, the value of P_{cu} would be more in multilayered winding. Though the thermal circuit of PET is coupled in a complex manner, the limiting values of B_m and J would be decided by the effectiveness of the thermal circuit [3,39,40]. With given loss, the safe operating temperature for the core would depend on the soft magnetic materials and that for copper would be on the insulation of litz wire strands as well as that placed between the layers.

Due to the complex nature of the thermal circuit, there exist multiple heat conduction channels with different heat transfer coefficients, and they are mostly coupled [3]. The directions of heat flow would be decided by the location of the hot spot temperature T_{hs} of the PET; it could be decided by the temperature differential $\Delta T = T_{hs} - T_{amb}$, such as,

$$T_{hs} - T_{amb} = \Delta T = R_{PET}(P_{core} + P_{cu}) \tag{13}$$

$R_{PET} = R_{th} + R_{conv}$ is the effective thermal resistance of the PET and T_{amb} is the ambient temperature. R_{PET} mostly consists of thermal conduction (R_{th}) and thermal convection (R_{conv}). Though it plays certain role, the heat transfer by radiation is ignored here. For compact design of PET, apart from reducing the total loss P_{tot}, the value of ΔT should be minimum. Large surface area of the core and the secondary winding are available for heat transfer. The major part of P_{tot} is removed by thermal convection [21,41,42] where the speed of the moving medium would play a significant role. The expressions of conductive and convective thermal resistances are,

$$R_{cond} = \frac{1}{h_{cond}A_{cond}} \text{ and } R_{conv} = \frac{1}{h_{conv}A_{conv}} \tag{14}$$

A_{cond} and A_{conv}, respectively, are areas available for conduction and convection and h_{cond} and h_{conv} are corresponding heat transfer coefficients. The value of h_{conv} depends on thermal conductivity of the attached medium and also on its speed where fan cooling improves its value significantly. Backed by practical validation, the finite element method (FEM) was extensively used to establish reasons behind the formation of hot spots in core and copper in high-power transformers [3]. It was realized that the temperature rise in copper was alarming on multilayer winding with constrained heat transfer features, e.g., the primary winding of Figure 3a. In multilayer winding, a significant part of P_{cu} is concentrated in the internal layer of primary winding closest to the core where the proximity effect is more prominent. The hot spot temperature T_{hs} is located here; its value needs to be reduced. It could be made possible if a part of P_{pri} close to I-section, in particular, is channelized to the ambience through the core. Considering the heat conduction is symmetrical around the I-section of Figure 3a, the overall heat conduction circuit of half of the PET (shown in Figure 6a), is represented in Figure 6b. The part P_{cu1} of total copper loss P_{cu} that could be channelized through the core is expressed as [41],

$$P_{\text{cu1}} = P_{\text{cu}} \frac{R_{\text{w}} + R_{\text{wa}} - R_{\text{ca}} \frac{P_{\text{core}}}{P_{\text{cu}}}}{R_{\text{w}} + R_{\text{wa}} + R_{\text{ca}} + R_{\text{F}}} = P_{\text{cu}} f(P_{\text{core}}, R_{\text{ca}}) \tag{15}$$

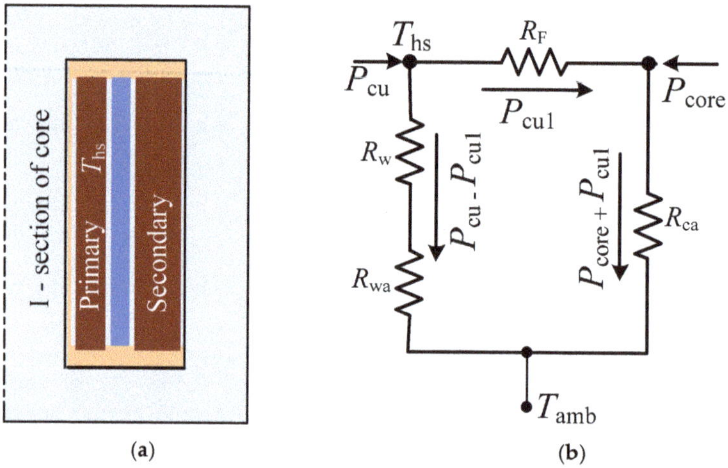

Figure 6. (**a**) Symmetrical half of PET of Figure 3a and (**b**) equivalent thermal circuit to remove heat loss.

R_{w} and R_{wa}, respectively, are the conductive and the convective resistance of the winding, and R_{F} and R_{ca}, respectively, are the effective thermal resistance of the coil former and the core. It is clear that more heat loss (i.e., P_{cu1}) would be channelized through the core if either of R_{ca} or P_{core} or together could be reduced through design or selection of suitable soft magnetic material; reduction of R_{F} would also play certain assisting role.

To achieve the abovementioned objective, it is proposed to use a mixed-core configuration for the magnetic circuit. It is known that a small fraction of core volume handles most of P_{tot} while the major part of its surface area is exposed to the ambience. Therefore, if that particular small section of core volume is replaced by a suitable soft magnetic material, then better heat removal feature by convection and radiation could be realized. The new core material is desired to possess following features:

1. Superior thermal conductivity.
2. Reduced core loss density.
3. Higher maximum operating temperature than the parent core.
4. The new material must be magnetically compatible.

Basic features of different soft magnetic materials are listed in Table 1.

3.2. Magnetic Compatibility of Different Types of Magnetic Circuits

For a given magneto motive force (MMF) Fm, the magnitude of magnetic flux φ linking the windings is decided by the reluctance value R_{m} of its magnetic circuit. The reluctance circuits of the core assemblies of commonly used high-power PETs of Figure 3a,b could, respectively, be represented in Figure 7a,b. Traditionally, the value of loss density P_{c} is considered at the same value everywhere. It means, for one type of cores, the value of B_{m} should be same everywhere. However, in the proposed idea of using the mixed core configuration, the value of loss density could be different. The expression of flux φ linking the windings is,

$$\varphi = \frac{F_{\text{m}}}{R_{\text{m}}} \quad \text{where} \quad R_m = \frac{l_{\text{m}}}{\mu_0 \mu_r A_c} \tag{16}$$

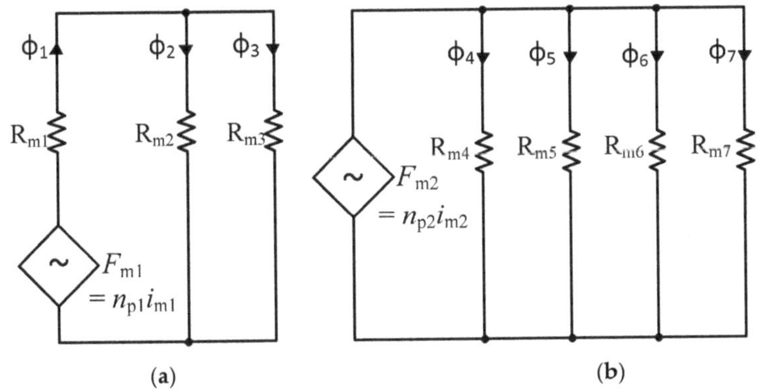

Figure 7. Reluctance circuit model of the (**a**) PET of Figure 3a and (**b**) PET of Figure 3b.

For Figure 7a, with MMF F_{m1}, the total flux $\varphi_1 = \varphi_2 + \varphi_3$ in the central limb is,

$$\text{or, } \varphi_1 = \frac{F_{m1}}{R_m} = \frac{n_p i_{m1}}{R_m}, \quad \text{where, } R_m = R_{m1} + \frac{R_{m2}R_{m3}}{R_{m2} + R_{m3}} \tag{17}$$

It is desired to have $\varphi_2 = \varphi_3$, and it could be achieved when $R_{m2} = R_{m3}$ is met.

For the magnetic circuit of Figure 7b, considering the same value of reluctance in each circuit, the total flux φ in the core assembly could be expressed as,

$$\varphi = \varphi_4 + \cdots + \varphi_7 \approx 4\frac{n_{p2}i_{m2}}{R_{m4}} \cdots \approx 4\frac{n_{p2}i_{m2}}{R_{m7}} \tag{18}$$

where i_{m2} is the magnetizing current and n_{p2} is the number of turns at primary. Neglecting the dimensional tolerance of the cores, the dynamic value of R_m of each parallel path plays an important role in flux distribution. When composite core segments are used, the dynamic behavior of μ_r becomes critical for designing the magnetic circuit of Figure 7b. The value of μ_r could change differently with respect to operating value of B_m, f_s, temperature, etc.

Characteristics of Different Soft Magnetic Materials

It is clear that for the magnetic compatibility, the role of relative permeability μ_r is extremely important. For better thermal performance, the heat distribution (Equation (15)) should be proper, i.e., more heat needs to be transferred to the core surface area exposed to the environment. Here, both the core loss density P_c and the thermal conductivity K would play significant roles. The basic parameters of popular soft magnetic materials are listed in Table 1 [20]. For ferrite cores, the parametric variation with respect to temperature [13] is large. Compared to others, its Curie temperature is much lower. The reduction in B_{sat} value vs. temperature is also sharp. For integration of different core types into a magnetic circuit of PET, the value of μ_r at different flux density and core temperature is important. The influence of temperature on μ_r of different soft magnetic materials is shown in Figure 8 [43] where nanocrystalline cores appear to be parametrically robust. It is also known that, for ferrite cores, the value of μ_r changes significantly with flux density [4]. However, for nanocrystalline cores, as shown in Figure 9, the value of μ_r is mostly constant, and only at large values of B_m is there gradual monotonic drooping in its value. Moreover, compared to other soft magnetic materials, the core loss of nanocrystalline cores at a particular frequency is comparatively less at any flux density (shown in Figure 10). Therefore, due to the parametric robustness, small core loss density, superior thermal conductivity, higher saturation flux density and high Curie temperature, the nanocrystalline cores of proper ribbon thickness are superior candidates for mixed-core configuration. It would not disturb the behavior of the magnetic circuit. If geometry permits, Fe-based nanocrystalline cores could ideally be suitable as flux integrators for medium- to high-frequency PETs. The drooping characteristics of μ_r, as shown in Figure 9, make these cores suitable for integration into a magnetic circuit where the MMF feeds several parallel magnetic circuits (shown in Figure 3b).

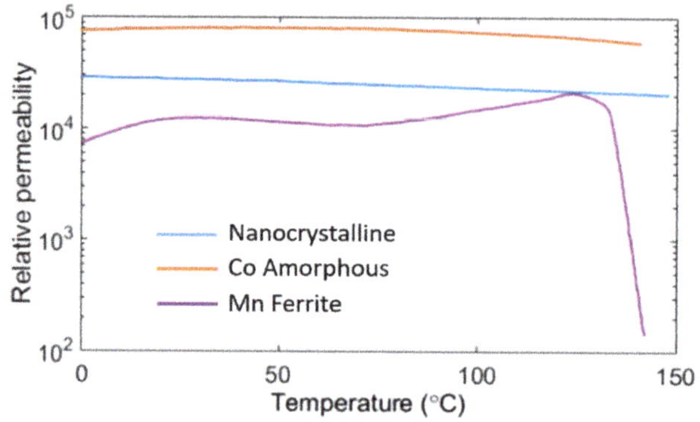

Figure 8. Behavior of relative permeability of soft magnetic materials vs. temperature [43].

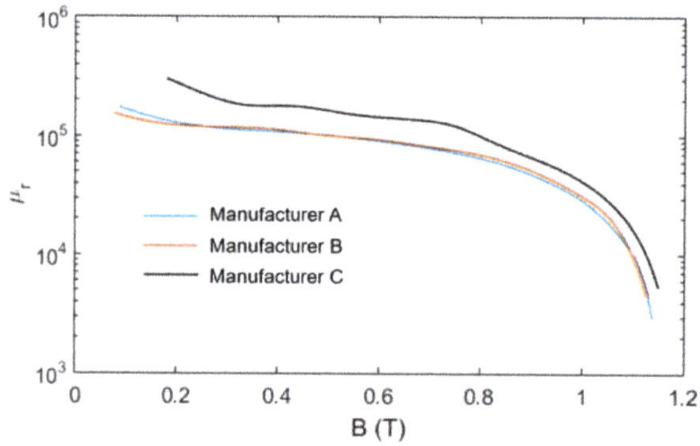

Figure 9. Permeability vs. flux density of competitive nanocrystalline cores have similar drooping characteristics.

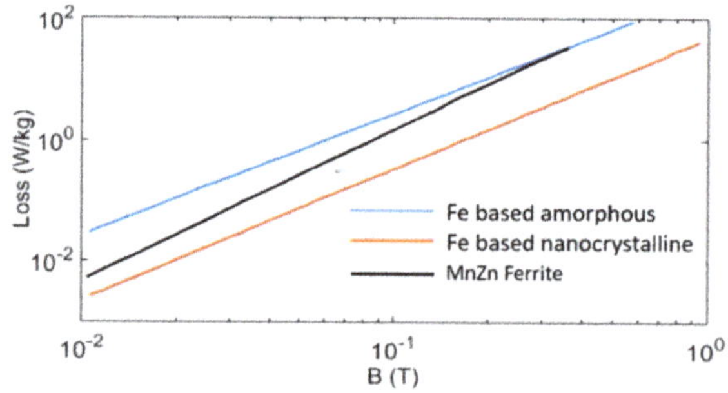

Figure 10. At a particular frequency the core loss density in nanocrystalline cores is minimum.

3.3. *Figure of Merit of Mixed-Core Magnetic Circuit for Series Reluctance Model*

It is known that for the PET of Figure 3a, the bulk of total power loss P_{tot} is concentrated around the central I-section EFGH. The hot spot temperature in the core and the primary winding, in particular, would reside in and around the I-section [3]. In order to properly utilize the large surface area of core exposed to the ambient medium for thermal convection and radiation, there needs to be an improvement in spreading the heat loss in the core by thermal conduction. A higher value of thermal conductivity K of core in the I-section EFGH would help remove the concentrated heat loss to the surface area of the magnetic circuit. As detailed in Equation (15), more power loss could be channelized if,

1. The value of P_c of the I-section is small.
2. The value of K of soft magnetic material used in the I-section is more.
3. The value of thermal resistance in coil former is reduced.

The first two features could simply be achieved if the central I-section is replaced by a geometrically compatible core material (shown in Figure 11a) with superior features so that the value of ΔT (see Equation (13)) in the I-section of core is reduced. The compatibility of the new material would be based on its dynamic magnetic parameters, core loss density P_c and the value of K. From the magnetic characteristics point of view, the role of I-section is not complicated. The *Figure of Merit* of the new core material for the I-section could be gauged by a suitability parameter S_{core}(SR); it is introduced in simple form as,

$$S_{core(SR)} = \frac{K}{P_c} \text{ and, } \mu_{rn} \geq \mu_{rp} \tag{19}$$

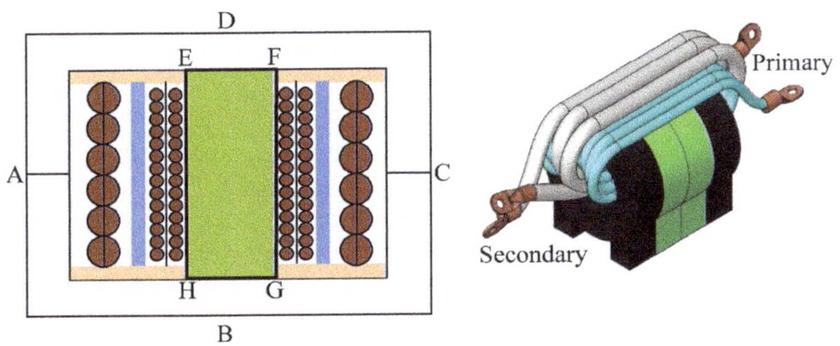

(**a**) (**b**)

Figure 11. Mixed-core assembly is for better use of the (**a**) electrical circuit of PET, i.e., the windings, and (**b**) magnetic circuit or the core assembly.

When compared with parent material, a smaller value of P_c and the same or larger value of K would result in superior distribution of heat in core. The permeability μ_{rn} of new I-section should be around the same value as μ_{rp}, i.e., of parent material. The value of K and other relevant parameters for different core materials are listed in Table 1.

3.4. *Figure of Merit of Mixed-Core for Parallelly Connected Magnetic Circuit*

Multiple toroidal cores are stacked (shown in Figure 3b) to support the desired flux level in the core of IHT. It is difficult to remove the heat loss from each core uniformly. The surface area of inner cores exposed to the ambient medium is relatively much less. It could lead to differential temperature rise across the core segments. In practice, the inner cores are found to be hotter. Like the PET of Figure 11a, here as well, as shown in the arrangement of Figure 11b, the problem could be reduced if the inner cores are replaced by a superior material. For a magnetic circuit where cores are used in parallel, it is difficult to establish the desired flux density everywhere. The dynamic value of relative permeability of each core being integrated would play an important role. To find the compatibility of the new core material for parallel reluctance model, and also to take care of sharing of flux, the suitability factor S_{core} of Equation (19) is modified to,

$$S_{\text{core(PR)}} = \frac{\mu_r(B_m)}{P_c} K, \quad \text{and for } B_{\text{m1}} > B_{\text{m2}}, \ \ \mu_{r1} < \mu_{r2} \tag{20}$$

where $\mu_r(B_m)$ is the value of μ_r of core at the operating value of B_m. It takes care of the distribution of flux when the condition (for $B_{m1} > B_{m2}$, $\mu_{r1} < \mu_{r2}$) is met. As shown in Figure 9, nanocrystalline cores support such characteristics. Furthermore, as shown in Figure 8, the value of μ_r for this core is also stable against temperature.

4. Experimental Validation of Mixed-Core Transformers

To validate the proposed idea of improving the thermal performance of a mixed-core PET, two divergent application domains were considered. In the first, to ignore any role of DC bias, the magnetic circuit was completed using two UU cores. In the second case, because the dynamics of B_m were sluggish, the ungapped toroidal shaped cores were assembled. In the first case, the prospect of transient DC bias was more. The magnetic circuit needed to be built with large DC bias capacity where, for large power applications, EE, UU or CC cores were suitable. On the other hand, for IHT, the prospect of transient DC bias was small. Therefore, ungapped toroidal cores with large permeability were preferred in the magnetic circuit.

4.1. Validation of Mixed-Core PET Where the Magnetic Circuit Uses UU cores

Ferrite cores in EE or UU shape are prominently used in high-power PET of Figure 3a where the central limb hosts both the windings. Due to their superior characteristics, as described in Equation (19), nanocrystalline cores would ideally be preferred for the I-section of Figure 11b. However, these ribbon-type cores are not dimensionally compatible as yet with the commonly available ferrite cores. It is difficult to match important dimensions one-to-one, e.g., the core area A_c, the window area A_w, the mean magnetic length l_m, etc.

It was difficult to procure I-shaped nanocrystalline cores physically or geometrically compatible to commonly used UU or EE ferrite cores. Therefore, to study the role of K and P_c on the thermal performance of PET, as shown in Figure 12, two mixed-core transformers (MCT) were built using different ferrite cores with different values of P_c and K. For core assembly, each MCT combined two different core types A and B procured from separate manufacturers. Relevant parameters of these cores are listed in Table 2.

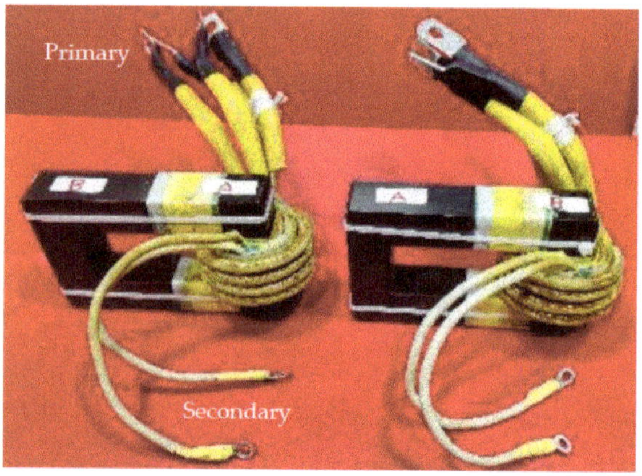

(**a**) (**b**)

Figure 12. Mixed-core transformers for FBDC where windings are overlaid on the core of material type (**a**) A for PET A and (**b**) B for PET B.

Table 2. Properties of ferrite cores type A and type B.

Core Material	Ferrite Core A		Ferrite Core B	
At temperature	25 °C	100 °C	25 °C	100 °C
Initial permeability	2500		2200	
B_{sat}, T	0.47	0.38	0.49	0.39
Curie temp., °C	>230		>210	
P_c @ 0.2 T, 25 kHz, kW/m^3	130	65	130	57
P_c @ 0.1 T, 100 kHz, kW/m^3	135	65	140	50
Thermal conductivity K, W/mK	≤4.3		5.0	

For testing of PETs A and B of Figure 12, one 20 kHz full-bridge DC−DC converter (shown in Figure 4) operating at 20 kHz was developed. The power controller is shown in Figure 13. For switching of inverter, IGBTs (Type: 2MBI075VAA-120-50) were used. The value of V_{DC} was 560 V. The value of inductor L1 was 100 µH. The winding layout of PETs A and B was similar, each had two secondary bifilar windings. To practically compare the magnetic compatibility at high flux density, the turns-ratio n_p:n_s:n_s of PETs was deliberately chosen at 24:2:2. The value of B_m at maximum value of d_{pwm} at 80% was 0.325T. Moreover, at the rated output power P_L at 4.5 kW (PWM duty cycle $d_{pwm} \approx 57\%$), the magnetic circuit would be loaded to the rated value of B_m at around 0.2 T. Each PET used one pair of UU cores, the value of A_c was 8.4 cm^2. At rated load current (I_a: 200 A) with d_{pwm} at 57%, the current density J (A/mm^2) at primary (strand dia.: 0.1 mm, 450 strands) and secondary litz wire (strand dia.: 0.1 mm, 3780 strands) conductors were 3.56 and 4.72, respectively. Due to higher value of J, the power loss in secondary conductors was more.

Figure 13. Prototype of a 4.5 kW DC−DC converter for performance testing of PET A and PET B.

Two sets of waveforms with an exactly similar nature of the magnetizing current i_{m1} (both secondary: open) in Figure 14a,b demonstrated that the core A and B were magnetically compatible even in the nonlinear zone of the B–H curve. At d_{pwm} of 80%, the value of B_m was deliberately kept large at 0.325 T so that the magnetic compatibility in the nonlinear zone was verified. The magnetic circuit had rated value of B_m (≈0.2 T) at designed power output.

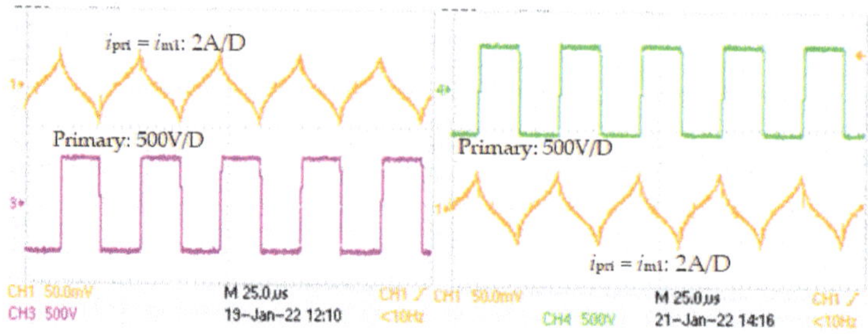

(**a**) (**b**)

Figure 14. At zero secondary current, even in nonlinear operating zone of B-H curve (B_m = 0.325 T), the behavior of the magnetic circuit of (**a**) PET A was similar to that of (**b**) PET B.

To evaluate the thermal performance comparatively, the two transformers were (resistive) load tested until the temperature in core and copper stabilized. The ambient temperature was 28 °C. The delivered power to load was 4.5 kW where the load voltage V_L was maintained at 22.5 V. At full load, the calculated value of B_m was 0.207 T. Various waveforms of the FBDC using PET A and PET B are shown in Figures 15a and 15b, respectively. Using the Fluke make 59 *Mini* IR thermometer, the temperature was recorded in each winding and also in the core segment close to the windings. The measured value of the hot spot temperature in the core and the windings of two PETs are listed in Table 3. It was clear that, compared to PET A, the temperature distribution of PET B was superior. The experimental results made it clear that the core type with lower value of P_c and/or possessing higher value of K would be more suitable for the core segment where the windings are laid. It could be stated that the situation would improve further if the I-section was replaced by a suitable nanocrystalline core material.

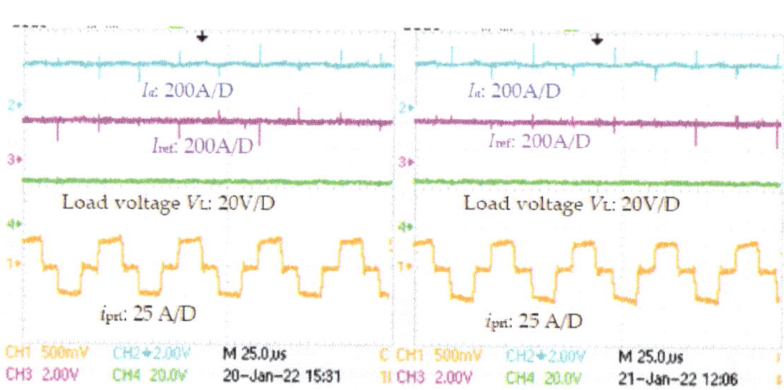

(**a**) (**b**)

Figure 15. Waveforms of FBDC controller when 200 A was drawn by the load and, in that loading condition, the magnetic circuit was loaded with B_m at 0.207 T, for both (**a**) PET A and (**b**) PET B.

Table 3. Measured hot spot temperature (°C) in core and windings (Ambient temp.: 28 °C).

Transformer	Primary	Secondary	Temperature of Core (°C)
PET A	86.9	92.8	Core A: 61.7
PET B	83.6	89.3	Core B: 57.3

The results detailed above were found to be reproducible when thermal images of PET A and PET B were captured by a camera (model: FLIR AX 5). In the captured images, the hot spot temperature of the secondary winding (SP1), primary winding (SP2) and also in the core (SP3) for two different case studies are shown in Figure 16 (for PET A) and in Figure 17 (for PET B). It was clear that the core of reduced value of P_c and of larger value of K helped reduce the hot spot temperature of PET B. The results mostly tallied with the findings of the temperature profile obtained through the noncontact type Fluke make 59 *Mini* IR thermometer.

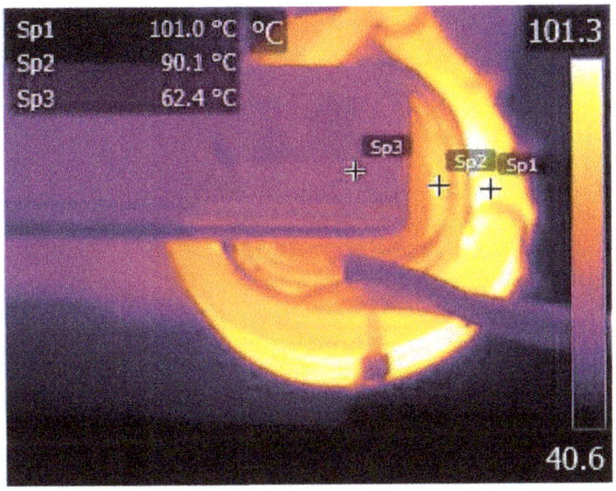

Figure 16. Temperature distribution of PET A using thermal camera (ambient temp.: 33 °C).

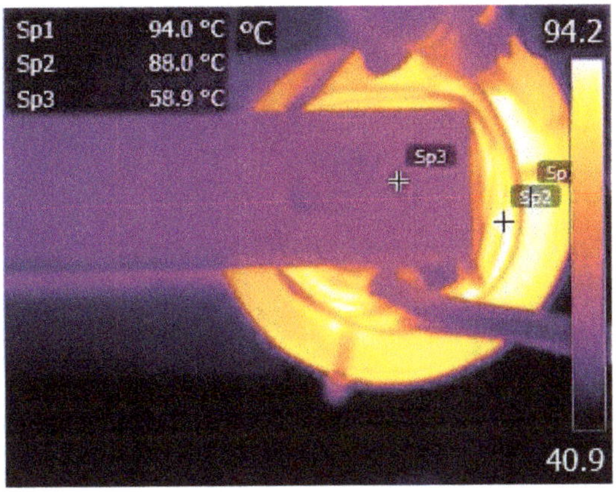

Figure 17. Temperature distribution of PET B using thermal camera (ambient temp.: 33 °C).

4.2. Validation of Mixed-Core IHT for Parallelly Connected Core Assembly

The loading pattern of the magnetic circuit of IHT is different. The value of P_{cu} is desired to be minimized because it is always at its rated value. It was analyzed in Section 2.2 that the ZVZCS topology along with using a low-loss magnetic circuit with a high effective value of μ_r and high saturation flux density B_{sat} would simultaneously minimize the copper content in IHT and also the value of P_{cu}. It requires the core to operate at a large value of B_m where cores with low loss density and high values of B_{sat} would be suitable. When both windings are placed in single layers, the impact of proximity effect on P_{cu} is minimized. Laterally, it would achieve better features for heat removal because windings are exposed to the ambient medium. For the magnetic circuit, the ungapped toroidal cores using nanocrystalline material would be a preferred choice because:

1. The value of magnetizing current would be small.

2. The value of leakage inductance would be negligible.
3. IHT does not need large DC bias capacity; transient load disturbance is small.
4. For large value of B_m (>0.25 T), ferrites are not suitable.

To practically study the usability of the proposed MCT, one induction heating controller was developed to deliver 40 kW output power. The complete experimental set up is shown in Figure 18. For comparative analysis, two transformers, i.e., the traditional single core-type IHT I and the mixed-core transformer IHT II, were designed using nanocrystalline cores. As shown in Figure 19a, in IHT I, four similar cores of, say, material type C, were stacked. Whereas in IHT II, as shown in Figure 19b, two poorly ventilated central cores were replaced by cores of different material, say, type D. The parametric details of both C and D type cores are listed in Table 4. Due to the reduced thickness of ribbons, when compared with core C, the value of P_c in core D was less. Naturally, the core D resulted superior value of S_{core}(PR) (see Equation (20)). The operating parameters of the inverter and cores are listed in Table 5. To study the magnetic compatibility, one single-turn search coil was wound on C and D cores each (shown in Figure 19b). The turns-ratio of each IHT was 8:3. For the tank circuit, the value of L4 was 52 μH and that of Cr was 3.6 μF. The inverter frequency was 12.5 kHz. With set coil current at 250 A, the value of J in primary (strand dia.: 0.1 mm, 2880 strands) and secondary (strand dia.: 0.1 mm, 2 × 3780 strands) conductors were 4.15 A/mm^2 and 4.21 A/mm^2, respectively.

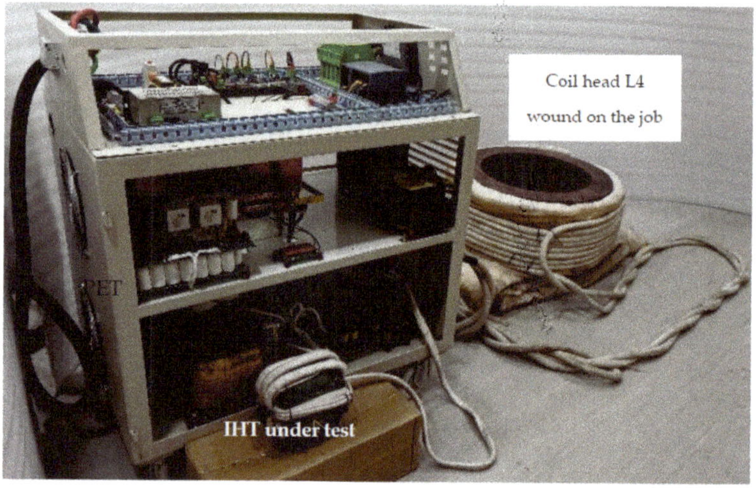

Figure 18. 40 kW induction heating controller for testing of IHTI and IHT II.

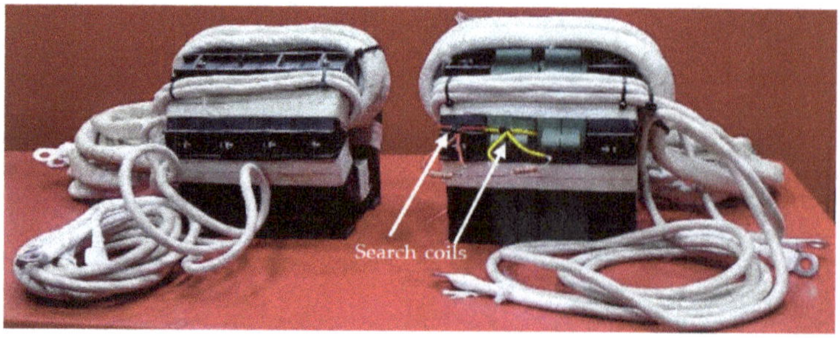

(a) (b)

Figure 19. Two 40 kW, 15 kHz transformers: (**a**) IHT I with original set of C cores and (**b**) IHT II with mixed-core configuration (C-D-D-C).

Table 4. Parameters of core *C* (black) and *D* (green).

Parameter	Core C	Core D
B_{sat} at 25 °C (130 °C), T	1.2 (1.18)	1.25 (1.21)
Initial permeability μ_r	>300 k	>300 k
K_{S1}, α, β	0.94, 1.4364, 1.638	0.22, 1.608, 1.681
Value of μ_r at B_m = 1 T	>12 k	>12 k
Ribbon thickness, μm	30	22
Area of each core, cm^2	5.25	5.62
Mean length of core, cm	29.8	29.8
P_c @0.2 T, 20 kHz, W/kg	4.98	1.82
P_c @0.6 T, 10 kHz, W/kg	11.2	3.78
Th. conductivity *K*, W/mK	10	10

Table 5. Operating parameters of IHT and core loss data.

	IHT I	IHT II
Delivered power, kW	35	
Input voltage to inverter at full load, V	450	
Inverter frequency at full load, kHz	12.5	
Core area, cm^2	21	21.74
Value of B_m in C, T	0.535	0.517
Value of B_m in D, T	-	0.517
Core loss density in each of C, W/kg	12.7	12.0
Core loss density in each of D, W/kg	-	4.21

Initially, the magnetic compatibility of cores C and D in the IHT II was tested for three different operating conditions; they are:

1. The magnetic circuit operated at maximum value of B_m when the secondary was kept open and the primary was excited with full voltage. The control circuit was disabled. The current at primary was the magnetizing current. Waveforms in Figure 21a validated that the two core types were magnetically compatible. Exactly similar nature of induced voltages in the single-turn search coils $V_{src\text{-}C}$ (core C) and $V_{src\text{-}D}$ (core D) proved that the flux density was shared appropriately.
2. The tank circuit was connected, but the power delivered through the coil head was zero. Moreover, there was a DC blocking capacitor C_{dc} of 100 μF added between primary of IHT II and the inverter output. The primary voltage was small. Here, as well, the two core types were found to be magnetically compatible (shown in Figure 21b) because even at very small flux density the readings in both the search coils were similar dynamically.
3. Magnetic compatibility of cores of IHT II was also tested when the secondary was loaded. The coil head L4 was loaded at 20 kW. The necessary waveforms are shown in Figure 20a. Here, as well, the voltage waveforms of both search coils appeared similar—in magnitude as well as in waveshape. Similar search coil voltage readings at zero secondary current as well as under loaded condition proved that cores C and D were integrated into the magnetic circuit of IHT II.

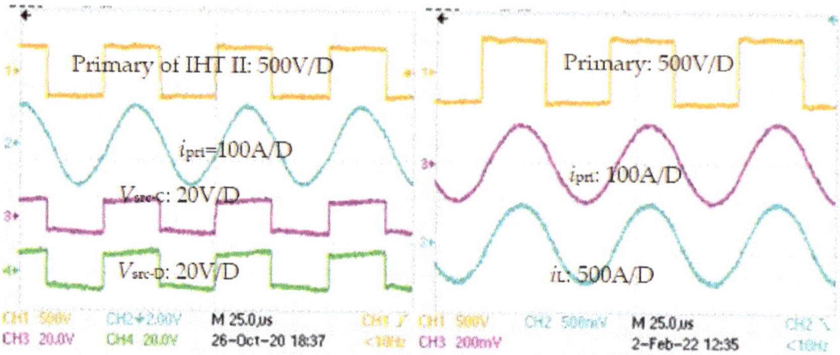

(**a**) (**b**)

Figure 20. (**a**) The magnetic compatibility of IHT II was also verified when the coil head was loaded. (**b**) Waveforms of different variables of the inverter when the IHT was loaded at 35 kW while delivering power to a section of pipe through the coil head L4.

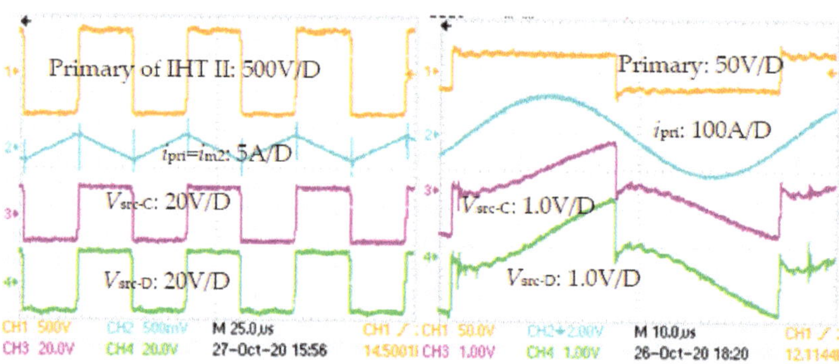

(**a**) (**b**)

Figure 21. Magnetic compatibility of C and D cores was verified through similar voltage readings in search coils wound on the respective cores, when (**a**) the secondary of IHT II was kept open, and (**b**) the DC blocking capacitor C_{dc} of 100 μF was added (shown in Figure 5) but the coil L4 was not loaded.

In order to gauge the gain of using the MCT, both IHT I and IHT II were put on a heat run test. The power drawn from the inverter was 35 kW (corresponding waveforms are shown in Figure 20b). IHTs were kept in open air under natural convection (shown in Figure 18). Each test was conducted until the steady state temperature in each core was attained. The ambient temperature was 26 °C. Using the noncontact type Fluke make 59 *Mini* IR and subsequently verified by the RTD thermometers at the end of the test, the temperature was recorded, as detailed in Figure 22, not only in each winding but also in each core segment. The measured hot spot temperature in different core and the windings of both the IHTs are shown in Figure 23. Due to the large value of K (see Table 4), the temperature distribution was found to be nearly uniform in each core segment. Moreover, compared to IHT I, temperature rise in each core of IHT II was reduced. The variation in temperature rise among different cores was also reduced. The temperature reading of internal cores with material D was much less when compared with that of IHT I. It essentially demonstrated that the power handling capacity of the MCT could be upgraded, and its power density could as well be increased.

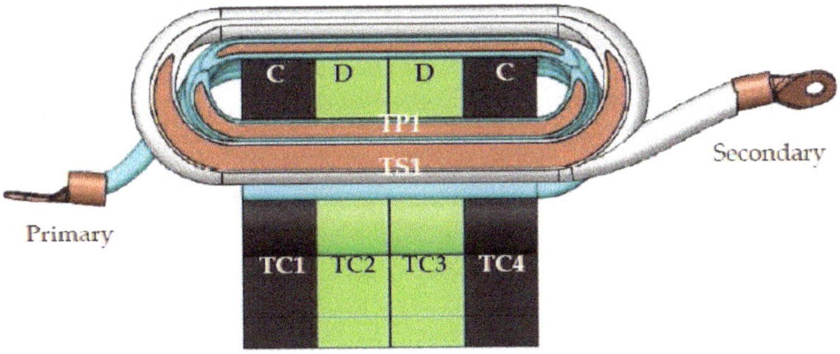

Figure 22. Sectional view of 40 kW transformer in mixed-core (core C and D) configuration.

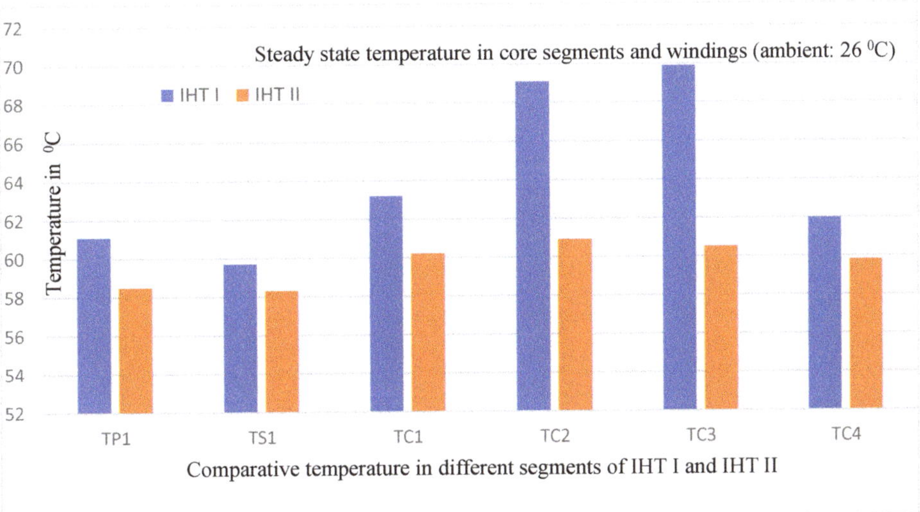

Figure 23. Compared to IHT I, the steady state temperature in different parts of IHT II was not only reduced, there was increased uniformity in maximum temperature in various parts of the mixed-core transformer.

5. Conclusions

The total power loss (i.e., core plus copper losses) in the majority of PETs would take place around a small core volume where the primary and secondary windings were overlaid. The surface area belonging to the rest of the core volume would remain exposed to the ambient environment. This article analyzed that the exposed surface area could be utilized effectively for heat transfer to the surrounding medium if the heat loss was easily channelized. The article proposed that a mixed-core configuration in the magnetic circuit could be used to achieve the goal. The idea was to replace the critical segment of the magnetic circuit having constrained heat conduction features by a geometrically compatible new core segment. The new segment would possess certain superior features, e.g., in core loss density, thermal conductivity, etc. To establish the compatibility of new core material, suitability factors were introduced for two different types of reluctance circuits. To validate the proposed idea, two transformers were built to cater two characteristically different applications. Initially, the magnetic compatibility of both the magnetic circuits under a worst-case operating condition were validated. Finally, both the transformers were put on prolonged load testing to comparatively validate their thermal performance. For both type of magnetic circuits, the superior thermal performance achieved during the heat run test validated the proposed idea. The reduced hot spot temperature in core and copper meant that the capacity rating of mixed-core transformer could be upgraded.

Funding: This research received no external funding and there is no APC involved for the article. The article is submitted on invitation.

Acknowledgments: The author thanks Vijay Lotekar, Vaibhav Arde, Mangesh Waman and Bhavna Rahate of Electronics Devices Pvt. Ltd. for assistance during experiments.

Conflicts of Interest: The author declares that there is no conflict of interest.

References

1. Popović-Gerber, J.; Oliver, J.A.; Cordero, N.; Harder, T.; Cobos, J.A.; Hayes, M.; O'Mathuna, S.C.; Prem, E. Power electronics enabling efficient energy usage: Energy savings potential and technological challenges. *IEEE Trans. Power Electrons* **2012**, *27*, 2338–2353. [CrossRef]
2. Hurley, W.G.; Wolfle, W.H.; Breslin, J.G. Optimized transformer design: Inclusive of high-frequency effects. *IEEE Trans. Power Electrons* **1998**, *13*, 651–659. [CrossRef]
3. Mogorovic, M.; Dujic, D. Thermal modeling and experimental verification of an air cooled medium frequency transformer. In Proceedings of the 2017 19th European Conference on Power Electronics and Applications (EPE'17 ECCE Europe), Warsaw, Poland, 11–14 September 2017; pp. 1–9.
4. Paul, A.K. ZVZCS SRI guides optimal use of copper and core for air-cooled nanocrystalline transformer for induction heating. *IEEE Trans. Ind. Appl.* **2020**, *56*, 970–978. [CrossRef]
5. Ruiz, R.D.; Venegas, R.V.; Anaya, R.A.; Moreno, G.E.; Rodríguez, R.J. Design and prototyping medium frequency transformers featuring a nanocrystalline core for DC−DC converter. *Energies* **2018**, *11*, 2081. [CrossRef]
6. Liu, J.; Sheng, L.; Shi, J.; Zhang, Z.; He, X. Design of high voltage, high power and high frequency transformer in LCC resonant converter. In Proceedings of the 2009 Twenty-Fourth Annual IEEE Applied Power Electronics Conference and Exposition, Washington, DC, USA, 15–19 February 2009; pp. 1034–1038.
7. Shen, W.; Wang, F.; Boroyevich, D.; Tipton, C.W., IV. High-Density Nanocrystalline Core Transformer for High-Power High-Frequency Resonant Converter. *IEEE Trans. Ind. Appl.* **2008**, *44*, 213–222. [CrossRef]
8. Yao, P.; Jiang, X.; Xue, P.; Li, S.; Lu, S.; Wang, F. Design optimization of medium-frequency transformer for DAB converters with DC bias capacity. *IEEE J. Emerg. Sel. Top. Power Electron.* **2021**, *9*, 5043–5054. [CrossRef]
9. Shih, L.; Liu, Y.; Chiu, H. A novel hybrid mode control for a phase-shift full-bridge converter featuring high efficiency over a full-load range. *IEEE Trans. Power Electron.* **2019**, *34*, 2794–2804. [CrossRef]
10. Costinett, D.; Seltzer, D.; Maksimovic, D.; Zane, R. Inherent volt-second balancing of magnetic devices in zero-voltage switched power converters. In Proceedings of the 2013 Twenty-Eighth Annual IEEE Applied Power Electronics Conference and Exposition (APEC), Long Beach, CA, USA, 17–21 March 2013; pp. 9–15. [CrossRef]
11. Choi, J.M.; Byen, B.J.; Lee, Y.J.; Han, D.H.; Kho, H.S.; Choe, G.H. Design of leakage inductance in resonant DC−DC converter for electric vehicle charger. *IEEE Trans. Magn.* **2012**, *48*, 4417–4420. [CrossRef]
12. Ram, B.S.; Paul, A.K.; Kulkarni, S.V. Soft magnetic materials and their applications in transformers. *J. Magn. Magn. Mater.* **2021**, *537*, 168210. [CrossRef]
13. Kauder, T.; Hameyer, K. Performance factor comparison of nanocrystalline, amorphous and crystalline soft-magnetic materials for medium frequency applications. *IEEE Trans. Magn.* **2017**, *53*, 8401504. [CrossRef]
14. Jimenez, H.O. AC resistance Evaluation of Foils, Round and Litz Conductors in Magnetic Components. Master's Thesis, Chalmers University of Technology, Göteborg, Sweden, 2013.
15. Morched, A.; Marti, L.; Ottevangers, J. A high frequency transformer model for the EMTP. *IEEE Trans. Power Deliv.* **1993**, *8*, 1615–1626. [CrossRef]
16. Mu, M. High Frequency Magnetic Core Loss Study. Ph.D. Thesis, Virginia Polytechnic Institute and State University, Blacksburg, VA, USA, 2013.
17. Muhlethaler, J.; Beila, J.; Kolar, J.W.; Ecklebe, A. Core losses under the DC bias condition based on Steinmetz parameters. *IEEE Trans. Power Electron.* **2012**, *27*, 953–963. [CrossRef]
18. Rodriguez-Sotelo, D.; Rodriguez-Licea, M.A.; Araujo-Vargas, I.; Prado-Olivarez, J.; Barranco-Gutiérrez, A.I.; Perez-Pinal, F.J. Power losses models for magnetic cores: A review. *Micromachines* **2022**, *13*, 418. [CrossRef] [PubMed]
19. Venkatachalam, K.; Sullivan, C.R.; Abdallah, T.; Tacca, H. Accurate prediction of ferrite core loss with nonsinusoidal waveforms using only Steinmetz parameters. In Proceedings of the 2002 IEEE Workshop on Computers in Power Electronics, 2002. Proceedings, Mayaguez, PR, USA, 3–4 June 2002; pp. 36–41. [CrossRef]
20. Wang, Y.R. Modelling and Characterization of Losses in Nanocrystalline Cores. Ph.D. Thesis, University of Manchester, Manchester, UK, 2015.
21. Bahamani, M. Design and Optimization Considerations of Medium-Frequency Power Transformer in High-Power DC−DC Applications. Ph.D. Thesis, Chalmers University of Technology, Gothenberg, Sweden, 2016.

22. Bahmani, M.A.; Thiringer, T. Accurate evaluation of leakage inductance in high-frequency transformers using an improved frequency-dependent expression. *IEEE Trans. Power Electron.* **2015**, *30*, 5738–5745. [CrossRef]
23. Hurley, W.G.; Gath, E.; Breslin, J.G. Optimizing the AC resistance of multilayer transformer windings with arbitrary current waveforms. *IEEE Trans. Power Electron.* **2000**, *15*, 369–376. [CrossRef]
24. Zhao, H.; Eldeeb, H.H.; Zhang, Y.; Zhang, D.; Zhan, Y.; Xu, G.; Mohammed, O.A. An improved core loss model of ferromagnetic materials considering high-frequency and nonsinusoidal supply. *IEEE Trans. Ind. Appl.* **2021**, *57*, 4336–4346. [CrossRef]
25. Li, Z.; Yao, K.; Li, D.; Ni, X.; Lu, Z. Core loss analysis of Finemet type nanocrystalline alloy ribbon with different thickness. *Prog. Nat. Sci. Mater. Int.* **2017**, *27*, 588–592. [CrossRef]
26. Zhou, Y.; Qi, B.; Zheng, M.; Cong, B. A novel DC bias suppression strategy for single-phase full-bridge DC−DC arc welding converter. *Electronics* **2021**, *10*, 428. [CrossRef]
27. Kohama, T.; Tokimatsu, S.; Shimamori, H. Elimination of magnetic saturation due to fast dynamic response in DC−DC converter. In Proceedings of the INTELEC 2009—31st International Telecommunications Energy Conference, Incheon, Korea, 18–22 October 2009; pp. 1–6. [CrossRef]
28. Paul, A.K. Choice of control function in magnetically-coupled full bridge DC−DC power controller for arc welding: A practical approach. *J. Power Electron. Devices Compon.* **2022**, *2*, 100005. [CrossRef]
29. Paul, A.K. Emulating full load testing of air-cooled nanocrystalline IHT at zero power. *IEEE J. Emerg. Sel. Top. Ind. Electron.* **2022**, *3*, 725–732. [CrossRef]
30. Wang, H.; Blaabjerg, F. Power electronics reliability: State of the art and outlook. *IEEE J. Emerg. Sel. Top. Power Electron.* **2021**, *9*, 6476–6493. [CrossRef]
31. Nikolov, G.T.; Valchev, V.C. Nanocrystalline magnetic materials vs ferrites in power electronics. *Procedia Earth Planetory Sci.* **2009**, *1*, 1357–1361. [CrossRef]
32. Hilal, A.; Raulet, M.A.; Martin, C.; Sixdenier, F. A comparative study: Dynamic and thermal behavior of nanocrystalline and powder magnetic materials in a power converter application. *J. Electron. Mater.* **2015**, *44*, 3768–3776. [CrossRef]
33. Paul, A.K. Robust product design using SOSM for control of shielded metal arc welding (SMAW) process. *IEEE Trans. Ind. Electron.* **2016**, *63*, 3717–3724. [CrossRef]
34. Malesani, L.; Mattavelli, P.; Rossetto, L.; Tenti, P.; Marin, W.; Pollmann, A. Electronic welder high frequency resonant converter. *IEEE Trans. Ind. Appl.* **1995**, *31*, 273–279. [CrossRef]
35. Ruiz-Robles, D.; Moreno-Goytia, E.L.; Venegas-Rebollar, V.; Salgado-Herrera, N.M. Power density maximization in medium frequency transformers by using their maximum flux density for DC–DC converters. *Electronics* **2020**, *9*, 470. [CrossRef]
36. Paul, A.K. Structured protection measures for better use of nanocrystalline cores in air-cooled medium-frequency transformer for induction heating. *IEEE Trans. Ind. Electron.* **2021**, *68*, 3898–3905. [CrossRef]
37. Meziane, B.; Zeroug, H. Improved efficiency determination for a PLL-controlled series resonant inverter for induction metal surface hardening. In Proceedings of the 2015 IEEE Industry Applications Society Annual Meeting, Addison, TX, USA, 18–22 October 2015; pp. 1–8.
38. Plesca, A. Considerations about maximum temperature of toroidal transformers in steady-state conditions. *J. Adv. Therm. Sci. Res.* **2020**, *7*, 22–29. [CrossRef]
39. Purushothaman, S.; de Leon, F. Heat-transfer model for toroidal transformers. *IEEE Trans. Power Deliv.* **2012**, *27*, 813–820. [CrossRef]
40. Lefevre, G.; Chazal, H.; Ferrieux, J.P.; Roudet, J. Application of Dovvell method for nanocrystalline toroid high frequency transformers. In Proceedings of the 2004 IEEE 35th Annual Power Electronics Specialists Conference (IEEE Cat. No.04CH37551), Aachen, Germany, 20–25 June 2004; Volume 2, pp. 899–904. [CrossRef]
41. Petkov, R. Optimum design of a high-power, high-frequency transformer. *IEEE Trans. Power Electron.* **1996**, *11*, 33–42. [CrossRef]
42. Jaritz, M.; Biela, J. Analytical model for the thermal resistance of windings consisting of solid or litz wire. In Proceedings of the 2013 15th European Conference on Power Electronics and Applications (EPE), Lille, France, 2–6 September 2013; pp. 1–10. [CrossRef]
43. Kącki, M.; Rylko, M.S.; Hayes, J.G.; Sullivan, C.R. Magnetic material selection for EMI filters. In Proceedings of the 2017 IEEE Energy Conversion Congress and Exposition (ECCE), Cincinnati, OH, USA, 1–5 October 2017; pp. 2350–2356. [CrossRef]

magnetism

Article

Magneto Elasticity Modeling for Stress Sensors

Gildas Diguet [1,2,*], Joerg Froemel [1,2], Hiroki Kurita [3], Fumio Narita [3], Kei Makabe [3] and Koichi Ohtaka [1]

1 Micro System Integration Center, Tohoku University, 519-1176 Aza Aoba, Aramaki Aoba-ku, Sendai 980-0845, Japan
2 Division for the Establishment of Frontier Sciences, Organization for Advanced Studies, Tohoku University, 2-1-1 Katahira, Aoba-ku, Sendai 980-8577, Japan
3 Department of Frontier Sciences for Advanced Environment, Graduate School of Environmental Studies, Tohoku University, Sendai 980-8579, Japan
* Correspondence: gildas.diguet.d4@tohoku.ac.jp

Abstract: In this article, the stress/stress sensing capability of FeSiB thin films is demonstrated and discussed. The sensing relies on the change in permeability by the application of stress, compressive and tensile, and the application of DC magnetic field. This susceptibility/permeability was tested by the exciting field (AC) being in the same direction with the applied stress. The susceptibility was shown to exhibit a maximal value at a given applied stress, the critical stress. Moreover, this maximal amplitude and position was changing with the application of an external DC magnetic field. For the DC field applied in the direction of the exciting field (AC) and longitudinal to the stress, the critical stress was shifted toward negative values and for the DC field applied perpendicularly, the critical stress was shifted toward larger positive values. This was experimentally demonstrated, and a model was constructed for a better understanding.

Keywords: magnetostriction; inductance; magnetic anisotropy; susceptibility; simulation; stress-impedance effect

1. Introduction

The magnetic properties of amorphous soft magnetic material can be used for magnetic sensors [1]. Typical commercial magnetic field sensors are based on the Giant Magneto Impedance (GMI) effect ($Z(H) - Z(H_{\max})/Z(H_{\max})$, with Z being the impedance, H the magnetic field and $H_{\max}$ the largest applied field. This can sense magnetic field down to $\mu_0 H = 10^{-12}$ T [1]. To produce this effect, an AC current is flowing through the material. In this case, depending on the frequency, the current density is concentrated close to the material edge (skin). The current density has a characteristic depth δ, called skin depth, that can be expressed, for a semi-infinite planar interface as [2]:

$$\delta = \sqrt{\frac{1}{\mu_0 \mu_r \pi f \gamma}}, \tag{1}$$

where $\mu_0 = 4\pi 10^{-7}$ H/m is the vacuum permeability, μ_r is the material relative permeability, γ is the material electric conductivity and f is the AC current frequency. Large permeability is reducing the skin depth, and the impedance of the conductor becomes higher; and the low permeability is increasing the skin depth, resulting in lower impedance. The application of an external magnetic field can then reduce the magnetic permeability $\mu_r(H)$. The GMI effect consists of the change in the measured impedance due to the application of a DC magnetic field [3,4].

Change in material permeability can also be achieved by applying another type of external stimulus. Especially, in the stress impedance (SI) effect, a strain or stress σ is applied to the magnetic material and a change in the magnetic permeability $\mu_r(H, \sigma)$ also changes the measured impedance [5–7]: $Z(\sigma) - Z(\sigma_{max})/Z(\sigma_{max})$, with Z the impedance, σ the magnetic field and σ_{max} the largest applied stress [8].

The SI effect has been used to successfully demonstrate strain sensing devices based with wires, ribbons and also thin films [5,9–15]. They showed generally much higher gauge factors than devices, using conventional effects (resistive, semiconductor). Beside general investigations as a strain gauge, also more specific applications, such as for structural health monitoring, pressure sensors, or force sensors have been proposed [15,16]. In a recent work, an amorphous FeSiB material was used to demonstrate the application for stress sensor [17]. A thin layer of this material was sputtered on top of a silicon beam. When stress was applied to the beam, the bending of the beam results in compressive stress, or tensile stress, in the magnetic layer, depending on the curvature of the beam. An external magnetic field was applied along the beam direction or perpendicularly. This revealed the complete map of the magneto elastic effect on the magnetic permeability. To complete this work, a theoretical description of the system should be provided.

Some models of the GMI are presented in the literature [18–22], including the effect of the stress on the GMI [23] to optimize the GMI because a low magnetostriction constant was experimentally yielding to a large GMI measurement [24]. Although those models were matching the experimental data, few models can be found for the susceptibility change with stress, and most of them are modeling only the case with an applied stress perpendicular to the anisotropy axis, and without external field [25,26].

In the present article, calculation of susceptibility is performed to understand the effect for many possible cases (compressive and tensile stress, different magnetic field directions) and to be able to predict permeability behavior with field and stress. Such results can be used to optimize sensors based on the magnetic susceptibility change.

2. Materials and Methods

Sample production and mechanical test were described in [17] and are briefly summarized here: the magnetic material, Ti/$Fe_{79}Si_7B_{14}$/Ti, was sputtered using a magnetron system (QAM-4-S, ULVAC, Inc., Tokyo, Japan) on a 300 µm thick Si wafer and the thicknesses were 20/500/20 nm, respectively. The titanium acted as adhesion layer and surface protection. The samples were cut in rectangle with length l = 40.4 mm and width w = 7.7 mm. A three-point bending test was applied to the sample and the applied force was recorded using a load cell (Xforce P, ZwickRoell Corporation, Ulm, Germany) with maximal force of 5 N. A coil was wrapped around the sample with the coil axis being parallel to the sample length. The coil was connected to an impedance analyzer (PSM1735, Newton4th Ltd., Leicester, the UK) for inductance measurements, at a frequency of 20 kHz. An electromagnet was used to generate an external magnetic field, along or perpendicular to the sample axis, with an amplitude of up to $\mu_0 H$ = 40 mT.

Magnetic characterizations, the M versus H loop, were performed in a BHV-50H Vibrating Sample Magnetometer (VSM) from Riken Denshi Co. (Tokyo, Japan). The applied field ranged from −1000 Oe to 1000 Oe (approximately − 80 kA/m to + 80 kA/m in S.I. units). From the magnetic loop the relative permeability was estimated as $\mu_r(H) = 1 + (dM(H)/dH)$.

3. Experimental Results

The inductance change is presented in Figure 1 for the external field in the direction of the sample and in Figure 2 for the external field applied perpendicularly [17].

Figure 1 presents the result of the inductance measurement of the coil, for the case where the applied field was in the sample axis. For the case where there was no applied field, $H_{//} = 0$, the compressive stress affected the inductance differently when compared to the tensile one. The curve was not symmetric with respect to the force. The curve was maximum at a critical force F_c, and the decrease was smooth for the force on the compression side and sharp on the tensile side of the curve. At a large applied force, F > 1 N, the inductance became constant at a value of L ~ 3.46 µH. As the magnetic field

$H_{//}$was increased, the initial inductance measured at zero force was decreasing to the same inductance value of $L \sim 3.46$ µH. In fact, the longitudinal field was (i) shifting the inductance versus force curve toward compressive force and (ii) reducing the inductance amplitude. Eventually, at larger fields, the inductances versus force curves were constant with values of $L \sim 3.46$ µH in the force measurement range.

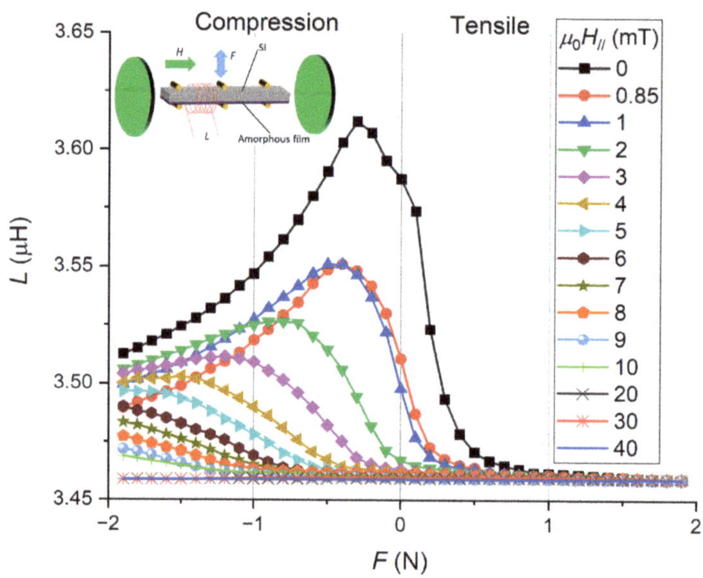

Figure 1. Measurement of the coil inductance versus the applied force with the applied longitudinal field.

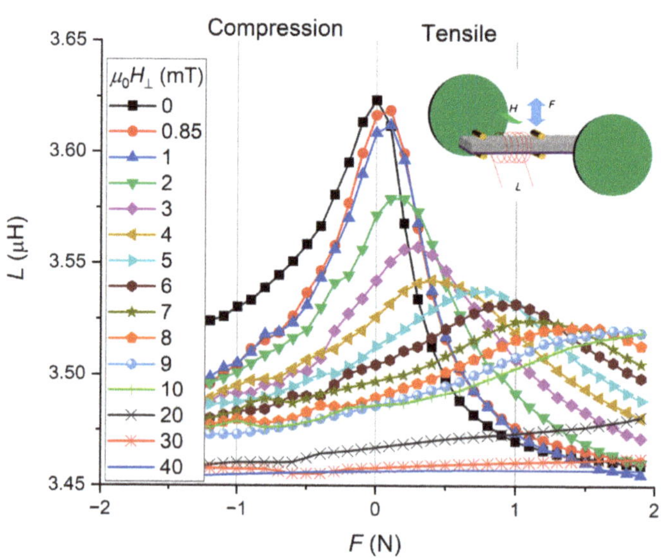

Figure 2. Measurement of the coil inductance versus the applied force with the applied orthogonal field.

After this set of measurements, performed in the longitudinal magnetic field configuration, the electromagnet was rotated to apply the magnetic field perpendicular to the sample axis. Figure 2 presents the result of the inductance measurement of the coil, for the case where the applied field was perpendicular to the sample axis. In this perpendicular configuration and for $H_\perp = 0$, the curve was different than the previous experiments in the longitudinal configuration and $H_{//} = 0$. The experiments were performed first, by applying force (compression and tensile) in the longitudinal configuration, then the experiments were performed by applying force (compression and tensile) after setting the configuration into the perpendicular configuration. Therefore, the sample has first experienced many stimuli (tensile and compressive stresses, plus longitudinal) large enough to affect the magnetic state. So, the state of these two curves, $H_{//} = 0$ and $H_\perp = 0$ has no reason to be exactly the same because of history. It is believed that there was a little hysteresis, but the shape of the curve remained very similar to the previous case: smooth for the force on the compression side and sharp on the tensile side of the curve. The effect of the field $H_\perp$ was different from the effect of $H_{//}$: the inductance peaks were shifting toward tensile force here. The inductances were reduced in both cases by the application of the magnetic field.

Figure 3a shows the result of the VSM measurement, and the magnetic loop is presented. Magnetic saturation was $M_S \sim 1.1$ MA/m and a coercive field H_c was ~ 400 A/m. The relative permeability was then extracted and plotted here versus the applied magnetic field in Figure 3a. This result was compared with the inductance measurement of the coil around the sample with the applied magnetic field and without an applied mechanical force.

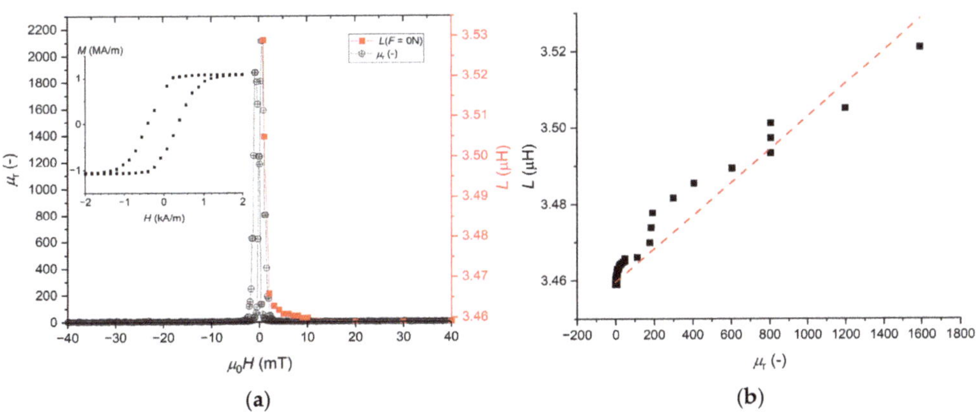

Figure 3. (**a**) Permeability obtained from the VSM (black circle) and coil inductance (red square) measured on a sample without a magnetic field; the inset shows the magnetic loop M versus H; (**b**) Approximation of the relationship between the measured inductance to the permeability.

High inductances were observed at high permeability values and lowest inductances are observed at lowest permeability values. In fact, when the magnetic field was larger than $\mu_0 H = 10$ mT, the permeability has a low value of $\mu_r = 1$; this was corresponding to a low and constant value of inductance $L \sim 3.46$ μH. This value $L \sim 3.46$ μH was then used to refer to the saturation of the sample. The inductance of a coil is usually expressed as $L = k\mu_r$ with the coefficient k being related to the number of turns of the coil, and the geometry of the coil and the sample [27,28]. The positive and negative sides of the $\mu_r(H)$ curve, Figure 3a, were then averaged as $\mu_r(H_i) = (\mu_r(+H_i) + \mu_r(-H_i))/2$ and plotted versus the coil inductances measured at the corresponding field $L(H_i)$. Figure 3b is showing the value of permeability plotted versus the inductance at the same magnetic field H_i amplitude: $\mu_r(H_i)$ versus $L(H_i)$.

A linear approximation of the relationship between permeability and inductance can be extracted as $L = (4.33 \times 10^{-5}) \times \mu_r + 3.46$. This line was also plotted in Figure 3b.

Figures 1 and 2 were presenting the inductance curve versus the applied force in the case of longitudinal DC field ($H_{//}$) and orthogonal DC field ($H_\perp$), respectively. The curves were presenting inductance peaks which can be characterized by the position of the peak, i.e., the critical force F_c, and the peak amplitude ΔL_{max}. This value of the peak amplitude ΔL_{max} was extracted by subtraction of the saturation value of 3.46 μH: $\Delta L_{max} = max(L) - 3.46\ \mu H$. This peak amplitude is an indicator of the permeability according to the relationship highlighted in Figure 3b as $\Delta L \sim \Delta \mu_r$.

Figure 4 shows the critical force F_c and the peak amplitude ΔL_{max} for all the tested configurations. The magnetic field direction was clearly changing the position of the inductance peak. Longitudinal field ($H_{//}$) was reducing the critical force ($dF_c/dH < 0$) and the orthogonal field ($H_\perp$) was increasing the critical force ($dF_c/dH > 0$). In both cases, a linear tendency was observed. The peak amplitudes ΔL_{max} were reduced by the magnetic field: with $H_{//}$, the decrease was faster than with $H_\perp$. The tendency of ΔL_{max} with applied fields was nonlinear.

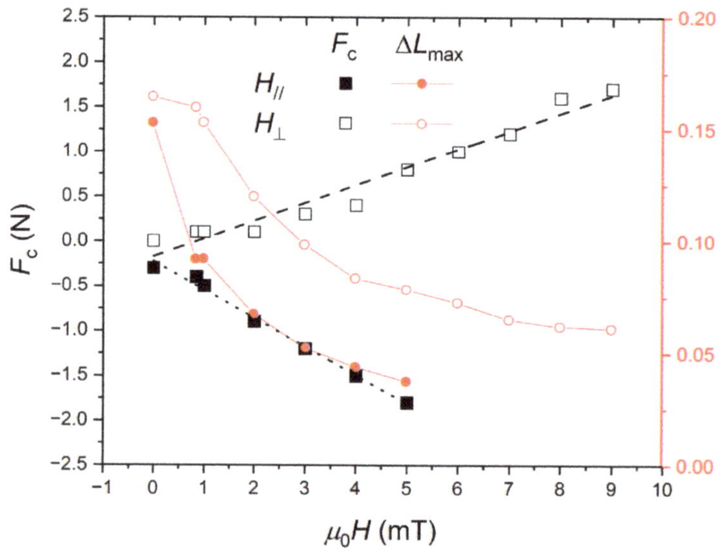

Figure 4. Experimental peak parameters: critical force F_c and maximal coil inductance change ΔL_{max} versus the applied force. Data for the longitudinal ($H_{//}$) and orthogonal ($H_\perp$) configuration.

4. Modeling

In this section, a model built to understand these experimental observations is described. According to the result of Figure 3b, the coil inductances can be interpreted by the magnetic state of the material. In this model, the magnetic states will be calculated as a function of stress and magnetic field. From these calculated magnetic states, the susceptibilities χ will be derived. Susceptibilities and permeabilities of materials with such large permeabilities will be supposed to be similar as for $\mu_r \sim 10^3$, then $\chi = \mu_r - 1 \sim \mu_r$.

First, the coupling between the magnetization and the applied magnetic fields has to be expressed. For this coupling, there are two contributions: one from the AC exciting field h_{ac}, created by the coil, and the second one is from the applied DC magnetic field H.

In the following, we will assume that the AC exciting field has a small amplitude compared to the field; $h_{ac} << H$. The coupling between an applied magnetic field and a magnetization vector, with an amplitude M_s, is expressed by the Zeeman energy. The Zeeman energies are expressed for the two contributions as:

$$E_{Zeeman\_AC} = -\mu_0 \vec{M_S} \cdot \vec{h_{ac}}, \quad (2)$$

and:

$$E_{Zeeman\_DC} = -\mu_0 \vec{M_S} \cdot \vec{H}. \quad (3)$$

It can be noticed that these Zeeman energies are minimized when the magnetization and the field are collinear.

Secondly, the coupling between the magnetization and the magnetic anisotropy has to be taken into account. In the following, we will assume two contributions: one from the intrinsic magneto crystalline anisotropy (small for this amorphous material) and a second one from the extrinsic magneto elastic coupling. For the magneto crystalline anisotropy, the energy is expressed, assuming a uniaxial anisotropy in the y-direction, as:

$$E_{\mathrm{a}} = K_{\mathrm{a}} \sin^2(\theta), \tag{4}$$

where K_{a} is the first-order magnetic anisotropy constant and θ is the angle between the magnetization vector and the anisotropy axis. This energy term minimization would require that $\theta = 0$, hence, the magnetization should be in the anisotropy direction.

When stress σ is applied to the magnetostrictive material, the corresponding magnetoelastic anisotropy can also be expressed by Equation (4), with the corresponding constant K_σ being expressed here as:

$$K_\sigma = \frac{3\lambda_{\mathrm{s}}\sigma}{2}, \tag{5}$$

In this Equation (5), λ_{s} is the magnetostriction constant of the material.

The final state of the magnetization, under magnetic field H, is obtained by minimizing the sum of all these energetic contributions. For instance, in the case where the uniaxial magnetic anisotropy and the applied field are orthogonal, the total magnetic interaction energy E is expressed, as:

$$E(\theta) = K_{\mathrm{a}} \sin^2(\theta) - \mu_0 M_{\mathrm{S}} H \cos\left(\frac{\pi}{2} - \theta\right), \tag{6}$$

By defining the anisotropy field H_{a} as:

$$H_{\mathrm{a}} = \frac{2K_{\mathrm{a}}}{\mu_0 M_{\mathrm{S}}}, \tag{7}$$

the two magnetization components, M_{L} (longitudinal, along with the field direction) and M_{T} (transverse to the field direction), can be expressed, for field $H < H_{\mathrm{a}}$, as:

$$\vec{M} = M_{\mathrm{L}}\vec{e_x} + M_{\mathrm{T}}\vec{e_y} = \frac{M_{\mathrm{S}}}{H_{\mathrm{a}}}\left[H\vec{e_x} + \sqrt{H_{\mathrm{a}}^2 - H^2}\vec{e_y}\right], \tag{8}$$

For field $H > H_{\mathrm{a}}$ the two components are then reduced to $M_{\mathrm{L}} = M_{\mathrm{S}}$ and $M_{\mathrm{T}} = 0$. These two curves are plotted in Figure 5a. For $H < H_{\mathrm{a}}$, M_{L} is linearly increasing up to M_{S} and remained $M_{\mathrm{L}} = M_{\mathrm{S}}$ for $H > H_{\mathrm{a}}$. For $H < H_{\mathrm{a}}$, M_{T} is decreasing from M_{S} down to 0 and remained $M_{\mathrm{T}} = 0$ for $H > H_{\mathrm{a}}$.

The corresponding relative susceptibility ($\mathrm{d}M/\mathrm{d}H$) components, χ_{L} and χ_{T}, are directly obtained for field $H < H_{\mathrm{a}}$ from Equation (8) as:

$$\chi_{\mathrm{L}} = \frac{M_{\mathrm{S}}}{H_{\mathrm{a}}}, \tag{9}$$

and:

$$\chi_{\mathrm{T}} = -\frac{M_{\mathrm{S}}}{H_{\mathrm{a}}}\frac{1}{\sqrt{\left(\frac{H_{\mathrm{a}}}{H}\right)^2 - 1}}, \tag{10}$$

For field $H > H_{\mathrm{a}}$ the two components are then reduced to $\chi_{\mathrm{L}} = 0$ and $\chi_{\mathrm{T}} = 0$. These susceptibilities can be normalized according to the constant $\chi_{\mathrm{a}} = M_{\mathrm{S}}/H_{\mathrm{a}}$.

These two curves are plotted in Figure 5b. It can be pointed out that, at $H = H_{\mathrm{a}}$, there was no susceptibility calculated. In fact, plotting the angle dependency of E for different values of field H could reveal this particularity, as explained in detail in Appendix A.

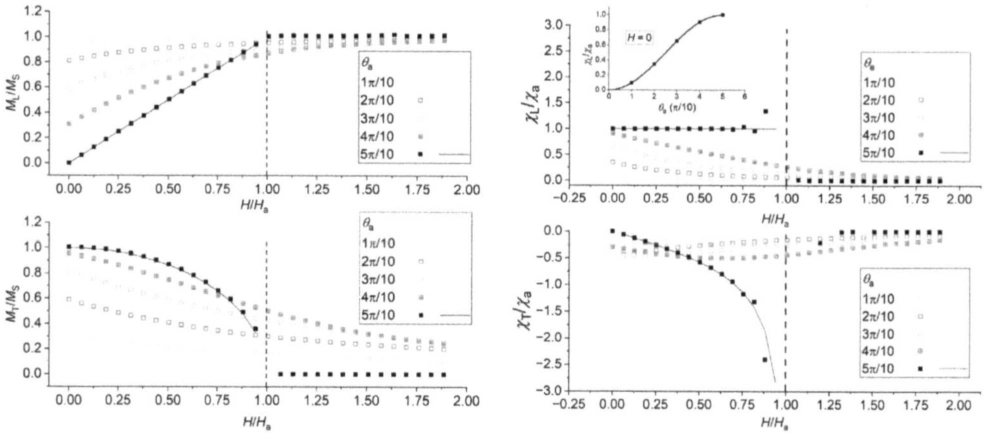

(a) (b)

Figure 5. (**a**) Relative magnetization components (M_L/M_S and M_T/M_S) versus H/H_a for different magnetic anisotropy direction θ_a; (**b**) Relative susceptibility components (χ_L/χ_a and χ_T/χ_a) versus H/H_a for different magnetic anisotropy direction θ_a. The full lines are theoretical values (Equation (8) for M_L and M_T, Equation (9) for χ_L and Equation (10) for χ_T) and square are numerical results from model.

5. Calculated Results of the Model

The model was used in COMSOL software to calculate the magnetization direction and then the susceptibility. This AC susceptibility was then calculated from the dependency of the magnetization with the AC field h_{ac}. It was assumed that h_{ac} is in the sample's main axis, the x-direction (longitudinal direction). The longitudinal direction is then referred below to as the x-direction and the orthogonal (perpendicular) direction as the y-direction. The susceptibility was extracted from the x-component of the magnetization as $\chi_L = dMx/dh_{ac}$. This susceptibility was then calculated with the two types of external stimuli: applied DC field H and stress σ. Especially, the DC field can be in the x-direction (longitudinal configuration: $H_{//}$) or in the y-direction (orthogonal configuration, $H_{\perp}$). In addition, the stress was supposed to be applied in the x-direction. Both tensile stress (positive values) and compressive stress (negative values) were considered. Susceptibilities are then plotted in the normalized unit (χ_L/χ_a) and the same for the applied field (H/H_a) and stress (K_σ/K_a) to provide a general overview of the phenomenon. A summary of direction of physical quantities is presented in Table 1.

Table 1. Physical quantity of the model.

Quantity	Direction
DC field H	Field $H_{//}$ (x-direction) or Field $H_{\perp}$ (y-direction)
AC field h_{ac}	x-direction
Stress σ	x-direction
Anisotropy axis	At angle θ_a to x-direction
Magnetization M	Longitudinal M_L (x-direction) or Transverse M_T (y-direction)

5.1. Without Magnetostriction

Figure 5a is presenting the curves M_L/M_S and M_T/M_S versus different anisotropy angles θ_a. The component M_L is linear with the applied field $H < H_a$ and is constant for the applied field $H > H_a$ for $\theta_a = \pi/2$. The corresponding susceptibility, presented in Figure 5b, is constant for $H < H_a$ as $\chi_L = \chi_a$ and for $H > H_a$ as $\chi_L = 0$. For other angles, the longitudinal component M_L is continuously increasing with H with an asymptotic behavior tending to the value M_S. The corresponding susceptibility χ_L is observed to continuously decrease with H tending toward the value 0. The inset of Figure 5b) is presenting the effect of θ_a on χ_L at $H = 0$: this is following the law $\chi_L = \chi_a \sin^2(\theta_a)$.

5.2. With Magnetostriction

5.2.1. No External Field

Figure 6 is presenting the curves χ_L/χ_a versus the anisotropy ratio K_σ/K_a: defined as the ratio between the magnetoelastic anisotropy (Equation (5)) and magnetic anisotropy constant, with an angle θ_a between these two anisotropy directions. The magnetostriction constant was supposed to be $\lambda_s = 30$ ppm. Firstly, it can be observed that at angle $\theta_a = \pi/2$, a peak appeared in the tensile stress part ($\sigma > 0$) at the exact value $K_\sigma = +K_a$. This situation is similar to the magnetic field $H = H_a$ applied perpendicularly to the anisotropy axis, as discussed previously, where the energy is independent of the magnetization angle (Appendices A and B), yielding to largest susceptibility. At angle $\theta_a = 0$, a sharp peak appeared in the compressive stress part ($\sigma < 0$) at the exact value $K_\sigma = -K_a$ because the total energy angular dependence would disappear.

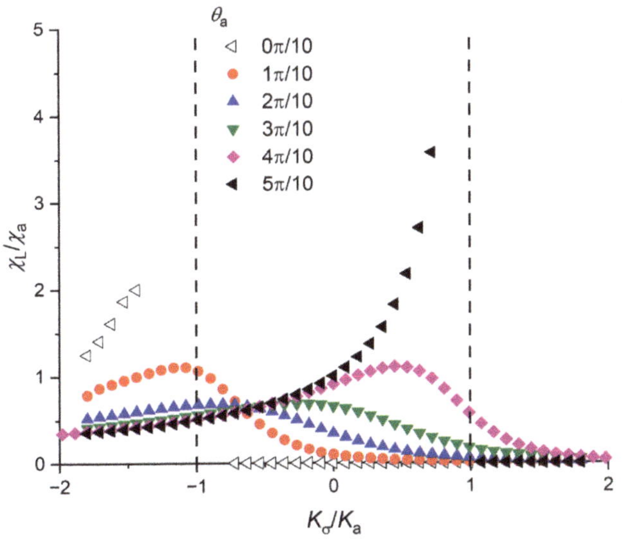

Figure 6. Relative susceptibility components (χ_L/χ_a) versus anisotropy ratio K_σ/K_a with $\lambda_s = 30$ ppm, for different angle θ_a between these two anisotropies: from 0 (same direction) to $\pi/2$ (perpendicular).

In the case $\theta_a = \pi/2$, the value $K_\sigma = +K_a$ was corresponding to $3\sigma\lambda_s = 2K_a$, defining the critical stress [25,26,29,30] as $\sigma_c = 2K_a/3\lambda_s$. The part of the curve with $\theta_a = \pi/2$ and $\sigma < \sigma_c$ was a smooth increasing curve, whereas for $\sigma > \sigma_c$ the curve became constant: the magnetic moments were now aligned with the stress. When the angle θ_a was reduced, the discontinuity disappeared and a continuous curve is observed until the angle was $\theta_a = 0$ where the curve was again discontinuous at stress value corresponding to $K_\sigma = -K_a$. For the intermediate angles, a bump appeared at an intermediate stress value ($-K_a \leq K_\sigma \leq K_a$). However, by reducing the angle toward $\theta_a = 0$, the global shape of the curve remained: below the critical stress ($\sigma < \sigma_c$) the curve was smoothly increasing and over the critical stress ($\sigma > \sigma_c$), the curve was more sharply decreasing.

The calculation was performed for different values of magnetostriction constant λ_s when no DC magnetic field was applied ($H_{DC} = 0$). The corresponding curves are plotted in Figure 7. It can be observed that the stress has no influence on the susceptibility for magnetostriction constant $\lambda_s = 0$. In this case, the susceptibility remains constant. For magnetostriction constant $\lambda_s > 0$, the susceptibility is then dependent on the amplitude of λ_s. For the lowest value, $\lambda_s = 5$ ppm, the susceptibility is increasing linearly with the applied stress in the range considered here. For the largest value $\lambda_s = 30$ ppm, the simulation was performed over a wider range of stress, revealing more about the behavior of the susceptibility curve: for large compression and large tensile stresses, the susceptibilities were very low whereas, at the critical stress σ_c, the curve is exhibiting a maximum value.

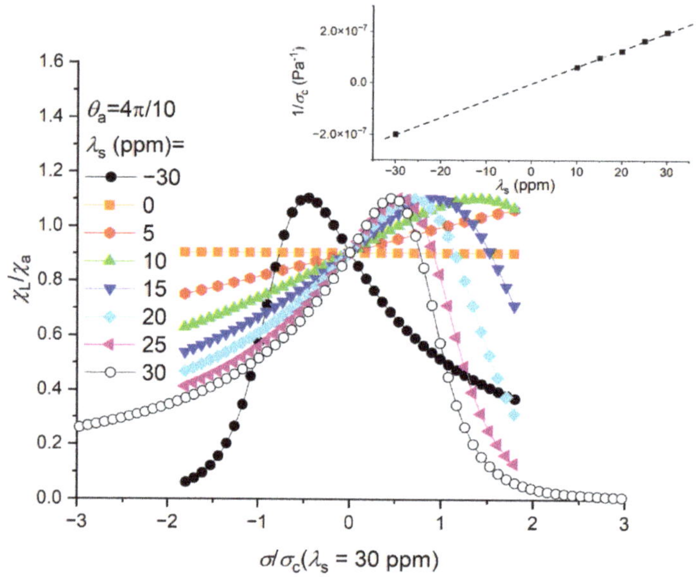

Figure 7. Relative susceptibility components (χ_L/χ_a) versus stress σ for different magnetostriction constant λ_s. The inset shows the inverse of the critical stress σ_c versus magnetostriction constant λ_s and dashed line is a guideline.

As the value of λ_s was increased, the critical stress σ_c was shifted to relatively smaller tensile stress. The inverse value of the extracted critical stress σ_c from these curves was linearly increased when the value of λ_s was increased. The susceptibility was defined, in the case of $\theta_a = \pi/2$, by Equation (7), as $\chi_L = M_S/H_a$, where H_a was defined in Equation (9). The susceptibility is $\chi_a = \mu_0 M_s^2/(2K_a)$ and if K_a is low, then is χ_a large. Adding the magnetostrictive term K_σ, then the effective anisotropy constant can be expressed as $K_a - K_\sigma$ (Appendix B). The largest susceptibility appeared at $K_a - K_\sigma = 0$, at the critical stress σ_c [25,26,29,30]. The critical stress depends on the magnetostriction constant as $\sigma_c = 2K_a/3\lambda_s \cong 1/\lambda_s$. For positive magnetostriction constant ($\lambda_s > 0$), the critical stress σ_c is positive (tensile) and for negative magnetostriction constant ($\lambda_s < 0$), the critical stress σ_c is negative (compression). The effect of positive and negative stress was also investigated here for value $\lambda_s = 30$ ppm and value $\lambda_s = -30$ ppm, as seen in Figure 7. The resulting curves were found to be symmetric to each other. The critical stress σ_c occurred at opposite values. This behavior was also observed in the literature [30,31]. For instance, Figure 8 shows an example of two materials permeability versus stress curves, one having a positive ($\lambda_s > 0$) and the other material with a negative magnetostriction constant ($\lambda_s < 0$). The two constants were not exactly opposite so the critical strains ($\varepsilon_c = E\sigma_c$, with E the material Young's modulus) were not opposite too. Nonetheless, the shapes of the two curves were in agreement with our calculations.

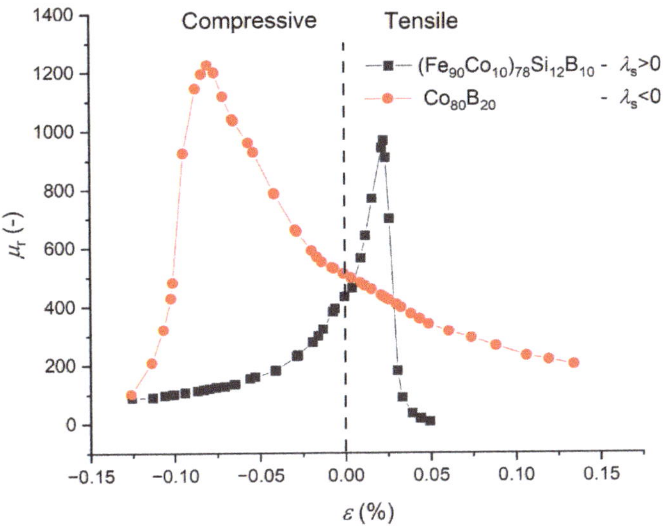

Figure 8. Permeability ($\mu_r \sim \chi$) versus strain ($\varepsilon = E\sigma$, with E the material Young's modulus of the materials) for a Fe-based material with positive λ_s and a Co-based material with negative λ_s [30].

5.2.2. With External Field

Finally, the results of the calculation, including the effect of the external magnetic field, are presented in this section. First, the effect of the longitudinal field is plotted in Figure 9. This curve is showing the susceptibility versus the applied stress and magnetic field applied in the same direction. The curves are observed to shift towards negative stress. These simulated curves are in agreement with the experimental curves presented in Figure 1. Magnetization is, without stress and field, directed along the magnetic anisotropy. As a stress is applied, the effective anisotropy is changing, as discussed in the previous: compressive stress is reinforcing a perpendicular anisotropy, and the whereas the tensile stress is lowering a perpendicular anisotropy. The effect of longitudinal field is similar to the tensile stress and the longitudinal magnetization component, and therefore the longitudinal susceptibility, are then affected (Figures A2 and A4).

The second configuration for the application of the external field, the perpendicular configuration, is presented in Figure 10. In this case, the effect was the opposite: the curves were shifted toward positive stress. The effect of perpendicular field is similar to the compressive stress which reinforced the effect of the magnetic anisotropy. The longitudinal magnetization component and therefore the longitudinal susceptibility are then affected (Figures A3 and A4).

Figure 11 is presenting the peak parameters extracted from these simulated susceptibility curves for comparison with the experimental results, presented in Figure 4. The behavior of the critical stress, in the longitudinal and perpendicular configurations, are matching the behavior of the critical force experimentally found for the same configuration in Figure 4: longitudinal field was linearly reducing the critical stress whereas the perpendicular field was linearly increasing this critical stress. From Figure 11, the slope of the critical stress σ_c versus applied field was extracted, by fitting the curve by linear fit method, for the case of longitudinal field ($H_{//}$) and for the case of perpendicular field ($H_{\perp}$). The results are presented in Table 2. The slope of the critical stress, in the longitudinal field configuration was of -1.72 whereas it was 1.27 in the perpendicular field configuration. The longitudinal value was 1.35 times larger than the perpendicular one (in absolute value). Experimentally, inductances were measured for the force applied during the bending test. This force is somehow believed to be proportional to the stress in the magnetic thin film. The change in the critical force measured under field is then proportional to the change in critical stress with magnetic field. From the experimental critical force, shown Figure 4, the same linear fit was performed to $F_c(H) - F_c(H = 0)$; the experimental offset was subtracted. The linear fit results are also presented in Table 2: in the longitudinal field configuration the slope was of -0.29 N/mT whereas it was 0.16 N/mT in the perpendicular field configuration. The experimental longitudinal slope value was 1.77 times larger than the perpendicular one (in absolute value). These two values, 1.35 (exp.) and 1.77 (sim.), are in relatively good agreement.

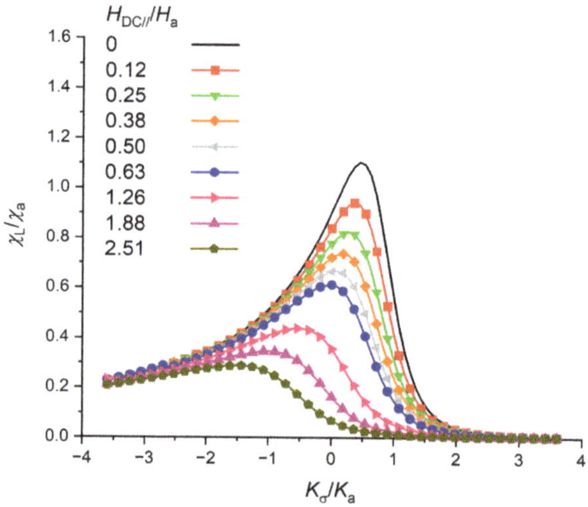

Figure 9. Relative susceptibility components (χ_L/χ_a) versus anisotropy ratio K_σ/K_a with λ_s = 30 ppm for different longitudinal field.

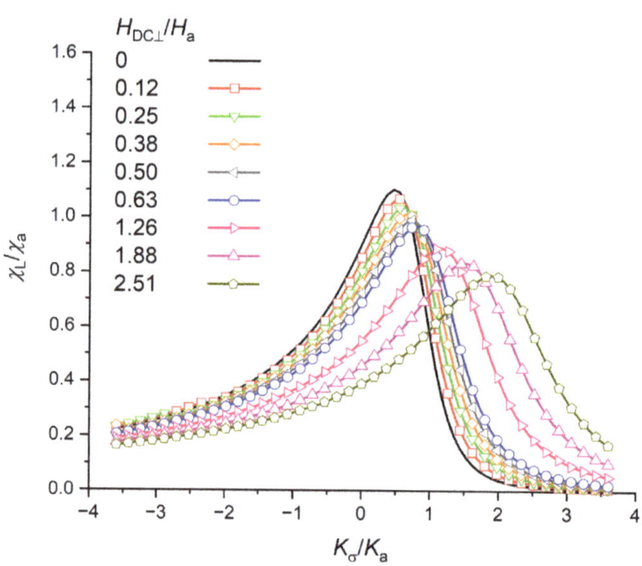

Figure 10. Relative susceptibility components (χ_L/χ_a) versus anisotropy ratio K_σ/K_a with λ_s = 30 ppm for different orthogonal field.

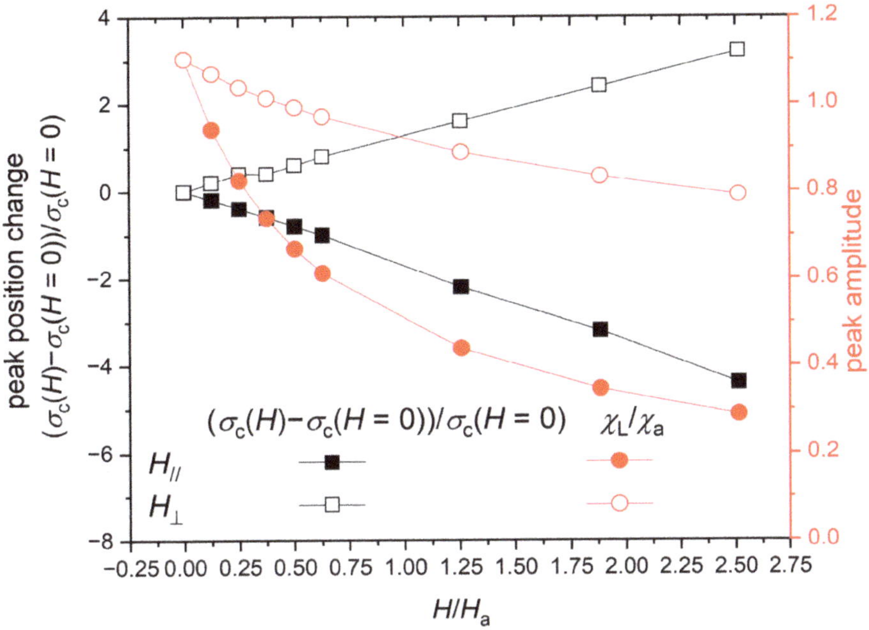

Figure 11. Peak parameters: critical stress σ_c and maximal susceptibility change versus the applied magnetic field.

Table 2. Fitting parameters for the critical stress and Force versus fields ($H_{//}$and $H_\perp$).

Quantity	Linear Slope Coefficient	Decay Coefficient
$[\sigma_c(H_{//}) - \sigma_c(H_{//}= 0)]/\sigma_c(H_{//} = 0)$	-1.72 (-) [1]	-
$[\sigma_c(H_\perp) - \sigma_c(H_\perp = 0)]/\sigma_c(H_\perp = 0)$	1.27 (-) [1]	-
$[F_c(H_{//}) - F_c(H_{//} = 0)]$	-0.29 N/(mT) [2]	-
$[F_c(H_\perp) - F_c(H_\perp = 0)]$	0.16 N/(mT) [2]	-
$\chi_L(H_{//})/\chi_a$	-	0.68 (-) [1]
$\chi_L(H_\perp)/\chi_a$	-	1.48 (-) [1]
$\Delta L_{max}(H_{//})$	-	1.32 (mT^{-1}) [2]
$\Delta L_{max}(H_\perp)$	-	3.67 (mT^{-1}) [2]

[1] Data extracted from Figure 11. Notice that parameters $[\sigma_c(H) - \sigma_c(H = 0)]/\sigma_c(H = 0)$, $\chi_L(H)/\chi_a$ and H/H_a have no units. [2] Data extracted from Figure 4.

The behavior of the simulated susceptibility peak amplitudes was also similar to the experimental results: the applied field was reducing the amplitude of the peak for the two configurations, although the decrease was faster in the case of the longitudinal configuration. To quantify these, a decay function: $y = A^*\exp(-H/H_{decay})$ was arbitrary selected to represented the behavior of the susceptibility peak amplitudes decrease with applied fields ($H_{//}$and $H_\perp$). The curves $\Delta L_{max}(H)$, obtained from Figure 4 for $H_{//}$and $H_\perp$, were fitted using this function. The results are given in Table 2. The decay coefficients H_{decay} obtained from the fitted curves of $\Delta L_{max}(H_{//})$ and $\Delta L_{max}(H_\perp)$ were then 1.32 mT^{-1} and 3.67 mT^{-1}, respectively. Hence, the experimental longitudinal coefficient value was 2.78-times smaller than the perpendicular one. The simulated $\chi_L(H_{//})/\chi_a$ and $\chi_L(H_\perp)/\chi_a$ versus H/H_a were fitted using the same decay function. The results are given in Table 2 as well. The decay coefficients H_{decay} obtained from the fitted curves of $\chi_L(H_{//})/\chi_a$ and $\chi_L(H_\perp)/\chi_a$ were then 0.68 and 1.48, respectively. Hence the simulated longitudinal coefficient value was 2.18-times smaller than the perpendicular one. These two values, 2.78 (exp.) and 2.18 (sim.), are in relatively good agreement.

Finally, the important parameters, for the stress sensor application [32], of such susceptibility versus stress curves are presented in Figure 12. The chosen conditions were selected to better understand the magnetization behavior in such materials, due to field and stress was classical material (FeSiB) for GMI and SI, but not optimized for the best condition. Stress-impedance sensors would have larger sensibility by several thermal/stress and magnetic treatments [3,33]. The model presented above might help to predict these parameters and to adapt and optimize the material's magnetic behavior, as a high value of susceptibility is needed to increase the sensitivity. The sensitivity direction can be selected to have a positive slope or negative slope by selecting the magnetostriction constant (Figure 7). In addition, the sensibility can be linear over a large range of stress by selecting a low amplitude magnetostriction constant, as seen in Figure 7.

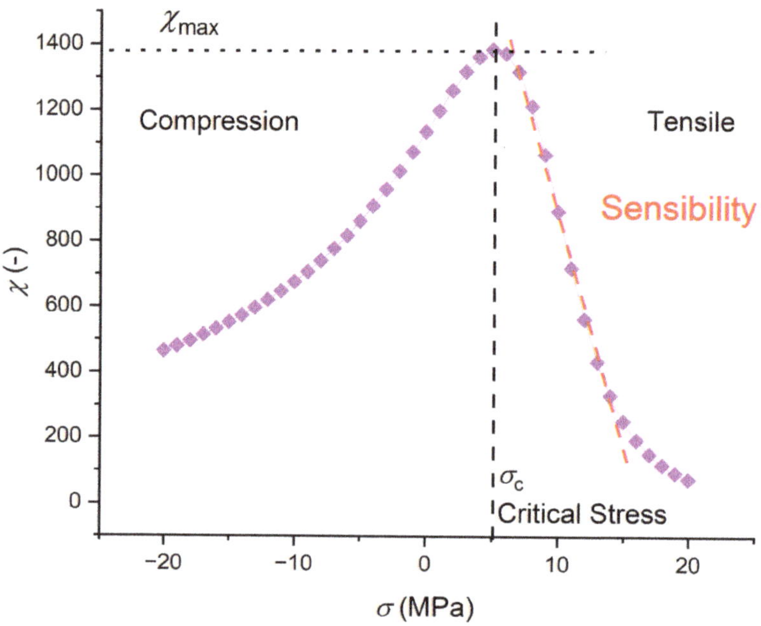

Figure 12. Stress sensor sensibility, calculated with M_S = 1 MA/m, H_a ~ 800 A/m, (χ_a = 1250) and with λ_s = 30 ppm and angle $\theta_a = 4\pi/10$.

If the shape of the μ_r (~χ) versus stress σ (~ε) was directly controlled by the sign of the magnetostriction constant λ_s, the position of the peak depends on the anisotropy stress direction as seen in Figure 6, which was observed in [6]. Moreover, this susceptibility peak can be shifted with a magnetic field, although this was decreasing the peak amplitude.

6. Conclusions

In this article, magneto elastic susceptibility was studied. Experimental data reported that when a force (stress) was applied in the same direction as the exciting field (AC field), for a critical force, the susceptibility was maximal. Under a longitudinal applied magnetic field, this critical force was reduced, even getting negative, and on the other hand, with the application of a perpendicular field, this critical force was shifted to larger values. In both cases, the observed peak amplitudes were reduced.

From these observations, a model of this magnetic behavior was constructed. This model assumed the magnetization rotation under the application of stress and magnetic field, and magnetic anisotropy. By setting the magnetostriction constant positive around

$\lambda_s = 30$ ppm, the shape of the susceptibility can be retrieved. Moreover, this model was successful to demonstrate similar behavior of the magnetic field effect on the susceptibility.

It is believed that this model is useful to provide a guideline in the conception of a stress sensor using the magneto elastic behavior of film based on the FeSiB type series; for stress impedance sensors or for coil-type sensors, as they are both sensitive to the magnetic state of the material under applied stress.

This can be used for both Fe-based ($\lambda_s > 0$) and Co-based ($\lambda_s < 0$) amorphous films. It can also help to predict the critical stress value, where the best sensitivity of the sensor can be found.

Author Contributions: Conceptualization, G.D., H.K. and J.F.; methodology, G.D., H.K., J.F. and F.N.; software, G.D.; investigation, K.M., G.D. and H.K.; writing—original draft preparation, G.D.; writing—review and editing, J.F., H.K., F.N., K.O. and F.N.; supervision, J.F. and F.N.; All authors have read and agreed to the published version of the manuscript.

Funding: This research was funded by Tohoku University Frontier Research in Duo grant, Japan Keirin Autorace foundation (JKA) and Ensemble Continuation Grant 2022.

Institutional Review Board Statement: Not applicable.

Informed Consent Statement: Not applicable.

Data Availability Statement: Not applicable.

Conflicts of Interest: The authors declare no conflict of interest. The funders had no role in the design of the study; in the collection, analyses, or interpretation of data; in the writing of the manuscript, or in the decision to publish the results.

Appendix A

These curves are presented in Figure A1. The curves corresponding to the case $H > H_a$ are clearly showing that the magnetization angle would be locked at an angle $\theta = \pi/2$: the field is strong enough to orientate the magnetization in the field direction. The curves corresponding to the case $H < H_a$ are showing a minimum between 0 and $\pi/2$: the magnetization is at an intermediate angle between the anisotropy axis and field direction. The curve corresponding to the case $H = H_a$ shows a broader minimum. This means that the magnetization can rotate more easily; the rotational susceptibility is then larger.

The Taylor expansion of E, Equation (6), around the angle $\theta = \pi/2$, is:

$$\widetilde{E}\left(\theta \sim \frac{\pi}{2}\right) \sim K - \mu_0 M_S H - \left(K - \frac{\mu_0 M_S H}{2}\right)\left(\theta - \frac{\pi}{2}\right)^2 + \cdots, \tag{A1}$$

Then, by rearranging the terms and using $H_a = 2K/\mu_0 M_s$, we obtain:

$$\frac{\widetilde{E}(\theta \sim \frac{\pi}{2})}{K} \sim 1 - \frac{2H}{H_a} - \left(1 - \frac{H}{H_a}\right)\left(\theta - \frac{\pi}{2}\right)^2 + \ldots, \tag{A2}$$

By setting the field value as $H = H_a$, the energy expression is finally simplified as:

$$\frac{\widetilde{E}(\theta \sim \frac{\pi}{2})}{K} \sim -1 + \cdots, \tag{A3}$$

This Expression (A3) shows that if the field is applied perpendicularly to the Easy Axis, and with amplitude as $H = H_a$, then the angular dependency vanished. These Taylor expansions, Equation (A1), are plotted for field values around H_a, showing the inversion of curvature as H is crossing the value $H = H_a$ which is a flat curve.

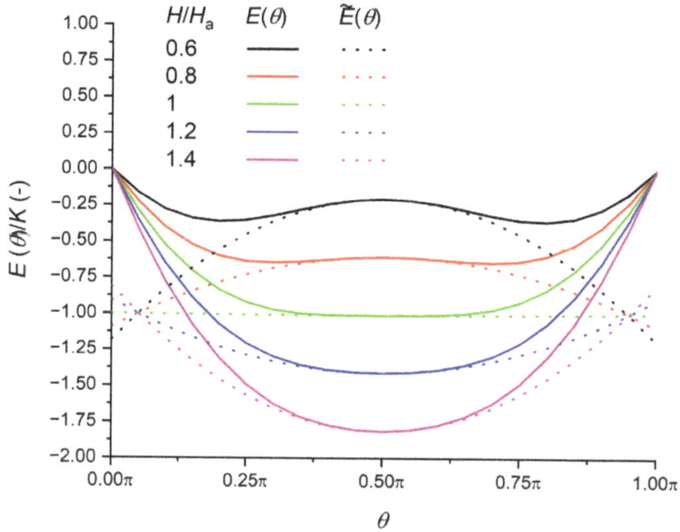

Figure A1. Angular dependence of the energy. Solid lines were constructed by using Equation (6) and dotted lines were based on the Taylor expansion, Equation (A3).

This was the simple case where a uniaxial magnetic anisotropy exists with $\theta_a = \pi/2$, at other angles, the magnetization curves were smoother and the susceptibility could be calculated for many field values. In the following, the effect of stress is considered along with the fields.

Appendix B

In the case of two uniaxial anisotropies, the treatment is provided in [34] and adapted below. Considering the anisotropy constants K_A (magneto crystalline) and $K_\sigma = 3\sigma\lambda_s/2$ (magneto elastic term) with perpendicular directions, then the total energy can be written as:

$$\begin{aligned} E(\theta) &= K_A \sin^2(\theta) + K_\sigma \sin^2(\tfrac{\pi}{2} - \theta) = K_A \sin^2(\theta) + K_\sigma \cos^2(\theta) \\ &= K_\sigma + (K_A - K_\sigma) sin^2(\theta) \end{aligned} \tag{A4}$$

If $K_A = K_\sigma$, the total energy is then independent of the angle. The first derivative of this energy is:

$$\frac{dE(\theta)}{d\theta} = (K_A - K_\sigma)\sin(2\theta), \tag{A5}$$

In addition, the second derivative of this energy is:

$$\frac{d^2E(\theta)}{d\theta^2} = 2(K_A - K_\sigma)\cos(2\theta), \tag{A6}$$

From Equation (A5), the minimum can be found at $\theta = 0$ and/or $\pi/2$; to determine whether the total energy is a minimum at these angles, the Equation (A6) must be positive. So, a condition on $(K_A - K_\sigma)$ appeared: if $K_A > K_\sigma$, then the minimum is for angle $\theta = 0$ (then M is along the magneto crystalline direction) and if $K_A < K_\sigma$, then the minimum is for angle $\theta = \pi/2$ (then M is along the stress direction). The term in $\sin^2(\theta)$, after adding the magneto elastic term, behaves with an effective constant $K_A - K_\sigma$. For tensile stress ($\sigma > 0$), the effective term in $\sin^2(\theta)$ is reduced whereas for compressive stress ($\sigma < 0$), the term in $\sin^2(\theta)$ is enhanced. Tensile stress is reducing the magneto crystalline anisotropy and compressive stress is reinforcing it.

Appendix C

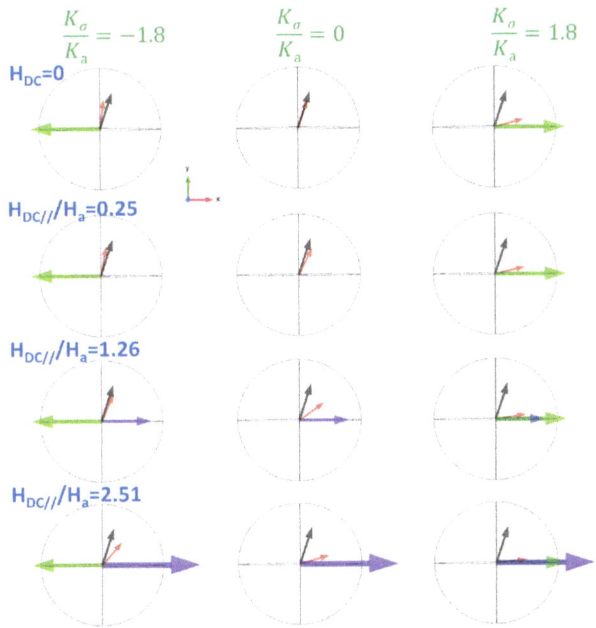

Figure A2. Snapshots of the magnetic state of the Magnetization (in red) at a given magnetic anisotropy (in black), for different stress (in green) and different longitudinal applied field (in blue).

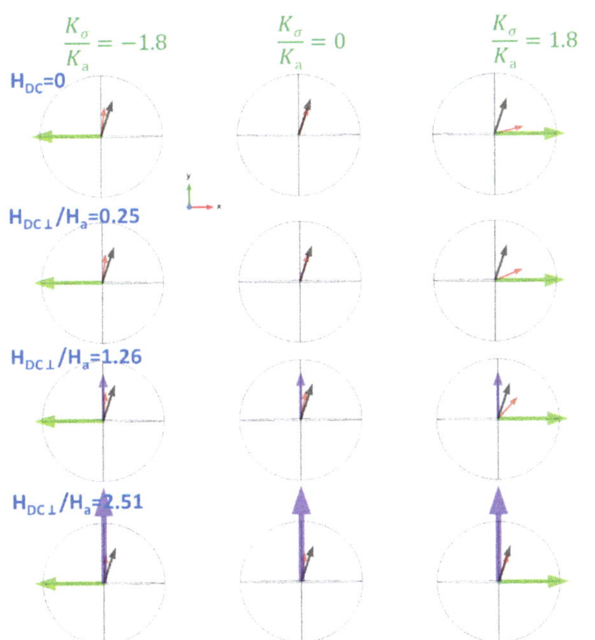

Figure A3. Snapshots of the magnetic state of the Magnetization (in red) at a given magnetic anisotropy (in black), for different stress (in green) and different perpendicular applied field (in blue).

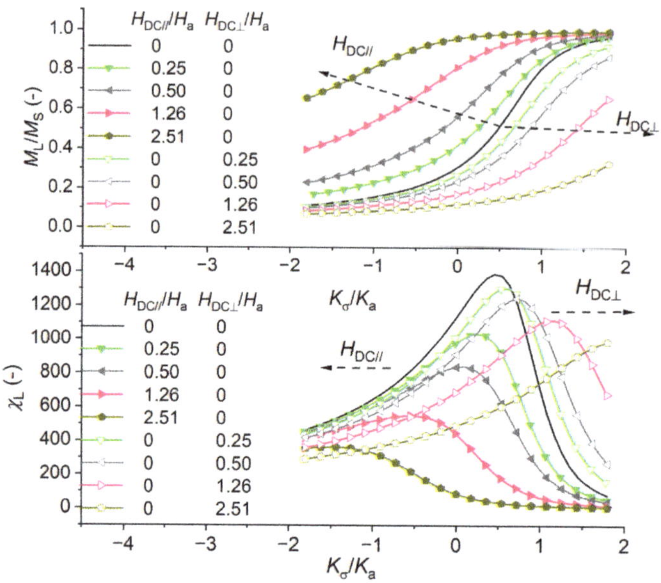

Figure A4. Relative longitudinal magnetization component (M_L/M_S) versus stress for different for different applied field (top) and corresponding relative longitudinal susceptibility component χ_L versus stress for different for different applied field (bottom).

References

1. Khan, M.A.; Sun, J.; Li, B.; Przybysz, A.; Kosel, J. Magnetic sensors-A review and recent technologies. *Eng. Res. Express* **2021**, *3*, 022005. [CrossRef]
2. Wheeler, H.A. Formulas for the skin-effect. *Proc. IRE* **1942**, *30*, 412–424. [CrossRef]
3. Garcia, D.A.; Munoz, J.L.; Kurlyandskaya, G.; Vasquez, M.; Ali, M.; Gibbs, M.R.J. Induced anisotropy, magnetic domain structure and magnetoimpedance effect in CoFeB amorphous thin films. *J. Magn. Magn. Mater.* **1999**, *191*, 339–344. [CrossRef]
4. Panina, L.V.; Mohri, K. MagnetoImpedance in Multilayer films. *Sens. Actuators* **2000**, *81*, 71–77. [CrossRef]
5. Fosalau, C.; Damian, C.; Zet, C. A high performance strain gage based on the stress impedance effect in magnetic amorphous wire. *Sens. Actuators A Phys.* **2013**, *191*, 105–110. [CrossRef]
6. Hashi, S.; Sora, D.; Ishiyama, K. strain and vibration sensor based on inverse magnetostriction of amorphous magnetostrictive films. *IEEE Magn. Lett.* **2019**, *10*, 8110604. [CrossRef]
7. Froemel, J.; Akita, S.; Tanaka, S. Simple Device to Measure Pressure Using the Stress Impedance Effect of Amorphous Soft Magnetic Thin Layer. *Micromachines* **2020**, *11*, 649. [CrossRef]
8. Blanco, J.M.; Zhokov, A.P.; Gonzalez, J. Torsional stress impedance and magneto-impedance in $(Co_{0.95}Fe_{0.05})_{72.5}Si_{12.5}B_{15}$ amorphous wire with helical induced anisotropy. *J. Phys. D Appl. Phys.* **1999**, *32*, 3140–3145. [CrossRef]
9. Shen, L.P.; Uchimiya, T.; Mohri, K.; Kita, E.; Bushida, K. Sensitive stress-impedance micro sensor using amorphous magnetostrictive wire. *IEEE Trans. Magn.* **1997**, *33*, 3355–3357. [CrossRef]
10. Mohri, K.; Uchimiya, T.; Shen, L.P.; Cai, C.M.; Panina, L.V.; Honkura, Y.; Yamamoto, M. Amorphous wire and CMOS IC-based sensitive micromagnetic sensors utilizing magnetoimpedance (MI) and stress-impedance (SI) effects. *IEEE Trans. Magn.* **2002**, *38*, 3063–3068. [CrossRef]
11. Tejedor, M.; Hernando, B.; Sanchez, M.; Prida, V.; Vasquez, M. Magneto-impedance effect in amorphous ribbons for stress sensor applications. *Sens. Actuators A Phys.* **2000**, *81*, 98–101. [CrossRef]
12. Garcia-Arribas, A.; Combarro, L.; Goiriena-Goikoetxea, M.; Kurlyandskaya, G.V.; Svalov, A.V.; Fernandez, E.; Orue, L.; Feuchtwanger, J. Thin-film Magnetoimpedance Structures onto Flexible Substrates as Deformation Sensors. *IEEE Trans. Magn.* **2017**, *53*, 2000605. [CrossRef]
13. Suwa, Y.; Agatsuma, S.; Hashi, S.; Ishiyama, K. Study on impedance change of strain sensor using magnetostrictive film. *J. Magn. Soc. Jpn.* **2010**, *34*, 342–346. [CrossRef]
14. Chen, J.A.; Ding, W.; Zhou, Y.; Cao, Y.; Zhou, Z.M.; Zhang, Z.M. Stress-Impedance effects in sandwiched FeCuNbCrSiB/Cu/FeCuNbCrSiB films. *Mater. Lett.* **2006**, *60*, 2554–2557. [CrossRef]
15. Vincent; Rodriguez, M.; Leong, Z.; Morley, N.A. Design and Development of Magnetostrictive Actuators and Sensors for Structural Health Monitoring. *Sensors* **2020**, *20*, 711. [CrossRef] [PubMed]

16. Froemel, J.; Diguet, G.; Muroyama, M. Micromechanical Force Sensor Using the Stress–Impedance Effect of Soft Magnetic FeCuNbSiB. *Sensors* **2021**, *21*, 7578. [CrossRef]
17. Kurita, H.; Diguet, G.; Froemel, J.; Narita, F. Stress sensor performance of sputtering Fe-Si-B alloy thin coating under tensile and bending loads. *Sens. Actuators A Phys.* **2022**, *343*, 113652. [CrossRef]
18. Usov, N.A.; Antonov, N.A.; Lagar'kov, A.N. Theory of giant magneto-impedance effect in amorphous wires with different types of magnetic anisotropy. *J. Magn. Magn. Mater.* **1998**, *185*, 159–173. [CrossRef]
19. Machado, F.L.A.; Rezende, S.M. A theoretical model for the giant magnetoimpendace in ribbons of amorphous soft-ferromagnetic alloys. *J. Appl. Phys.* **1996**, *79*, 6558. [CrossRef]
20. Jin, F.; Li, J.; Zhou, L.; Peng, J.; Chen, H. Simulation of Giant Magnetic Impedance effect in Co-based Amorphous films with demagnetizing field. *IEEE Trans. Magn.* **2015**, *51*, 7100404. [CrossRef]
21. Chen, H.; Britel, M.R.; Zhukova, V.; Zhukov, A.; Dominguez, L.; Chizik, A.B.; Blanco, J.M.; Gonzalez, J. Influence of AC magnetic field amplitude on the surface magnetoimpedance tensor in amorphous wire with helical magnetic anisotropy. *IEEE Trans. Magn.* **2004**, *40*, 3368–3377. [CrossRef]
22. Kraus, L. Theory of giant magneto-impedance in the planar conductor with uniaxial magnetic anisotropy. *J. Magn. Magn. Mater.* **1999**, *195*, 764–778. [CrossRef]
23. Peng, B.; Zhang, W.L.; Zhang, W.X.; Jiang, H.C.; Yang, S.A. Simulation of stress impedance effect in magnetoelastic films. *J. Magn. Magn. Mater.* **2005**, *288*, 326–330. [CrossRef]
24. Sartorelli, M.L.; Knobel, M. Giant magneto-impedance and its relaxation in Co-Fe-Si-B amorphous ribbons. *Appl. Phys. Lett.* **1997**, *71*, 2208. [CrossRef]
25. Spano, M.L.; Hathaway, K.B.; Savage, H.T. Magnetostriction and magnetic anisotropy of field annealed Metglas 2605 alloys via M-H loop measurements under stress. *J. Appl. Phys.* **1982**, *53*, 2667. [CrossRef]
26. Livingston, J. Magnetomechanical properties of amorphous metals. *Phys. Status Solidi A* **1982**, *70*, 591–596. [CrossRef]
27. He, D.F.; Shiwa, M. A magnetic sensor with amorphous wire. *Sensors* **2014**, *14*, 10644–10649. [CrossRef]
28. Mendoza Zelis, P. Magnetostrictive bimagnetic trilayer ribbons for temperature sensing. *J. Appl. Phys.* **2006**, *101*, 034507. [CrossRef]
29. Shin, K.H.; Inoue, M.; Arai, K.I. Elastically coupled magneto-electric elements with highly magnetostrictive amorphous films and PZT subtrates. *Smart Mater. Struct.* **2000**, *9*, 357–361. [CrossRef]
30. Ludwig, A.; Frommberger, M.; Tewes, M.; Quandt, E. High-frequency magnetoelastic multilayer thin films and applications. *IEEE Trans. Magn.* **2003**, *39*, 3062. [CrossRef]
31. Appino, C.; Fiorillo, F.; Maraner, A. initial susceptibility versus applied stress in amorphous alloys with positive and negative magnetostriction. *IEEE Trans. Magn.* **1993**, *29*, 3469–3471. [CrossRef]
32. Kubo, Y.; Hashi, S.; Yokoi, H.; Arai, K.; Ishiyama, K. Development of strain and vibration sensor using magnetostriction of magnetic thin film. *IEEE Trans. Sens. Micromach.* **2018**, *138*, 153–158. (In Japanese) [CrossRef]
33. Ding, Q.; Li, J.; Zhang, R.; He, A.; Dong, Y.; Sun, Y.; Zheng, J.; Li, X.; Liu, X. Effect of transverse magnetic field annealing on the magnetic properties and microstructure of FeSiBNbCuP nanocrystalline alloys. *J. Magn. Magn. Mater.* **2022**, *560*, 169628. [CrossRef]
34. Cullity, B.D.; Graham, C.D. Magnetic anisotropy. In *Introduction to Magnetic Materials*, 2nd ed.; John Wiley and Sons Inc.: Hoboken, NJ, USA, 2008.

magnetism

MDPI

Article

Vector-Based Magnetic Circuit Modelling of Induction Motors

Braden Kidd

Faculty of Information Technology, Monash University, 20 Exhibition Walk, Clayton, VIC 3800, Australia; braden.kidd@monash.edu

Abstract: Electro-mechanical devices incorporating rotating magnetic fields can be modelled using a wide range of analytical techniques. Choosing a modelling technique usually requires a trade off between computational efficiency and accuracy. Magnetic flux-based models aim to achieve an optimum balance between computational intensity and accuracy, as required for real time control applications. This paper demonstrates how vector-based magnetic circuit equations can be used to describe the operational characteristics of an induction motor at a more fundamental level than commonly used magnetic flux models. Doing so allows for closed form equations to be derived directly from device-specific geometry. The resultant model has advantages of numerical method-based analytical techniques while retaining the computational efficiency of closed form equations.

Keywords: magnetic circuit; rotating magnetic fields; direct quadrate model; magnetic equivalent circuits; induction motor

1. Introduction

Modelling electromagnetic devices can be achieved using simple analytical techniques such as those based on electrical and magnetic equivalent circuits. The elementary circuit elements upon which these equivalent circuits are based were originally derived to describe the characteristics of electromagnetic devices with no relative constituent motion. As such, using them to model devices incorporating rotational motion often requires a high degree of abstraction.

In the case of electrical equivalent circuits such as the Steinmetz equivalent circuit of an induction motor, parameters can either be empirically derived [1,2], or derived using more complex modelling techniques [3]. This results in a black box model, where internal workings and parameters are unavailable. The models' computational efficiency and the ability to extrapolate its usage to non-ideal operating conditions [4] has kept it relevant despite the availability of alternatives. However, the lack of access to internal dynamic parameters limits its ability to be extrapolated to describe more complex operational behaviour in its basic form [5].

A more fundamental approach to modelling electrical machines can be achieved using numerical method-based computer simulations at the expense of greater computational complexity. These modelling techniques divide the internal geometry of the simulated device into discrete elements, each with their own unique electrical, magnetic and mechanical properties. Common examples of this approach are variations of the Lumped Parameter Model (LPM) [6]. These models use an interconnected mesh of magnetic circuit elements to calculate magnetic flux. In its most basic application, only linear magnetic field behaviour can be modelled. However, its application can be extended to include modelling non-linear behaviour such as magnetic saturation [7], fault detection [8] and thermal analysis [9].

The Finite Element Model (FEM) is another commonly used numerical method technique. It works by solving partial differential equations derived directly from Maxwell's equations at the boundaries of each discrete element. This is a more fundamental approach to modelling electrical and magnetic fields compared to LPM and is therefore regarded as being more accurate [10,11]. A trade off for this improved accuracy is increased computation time [12]. Hybrid FEM models also exist that reduce computational complexity using either equivalent circuits [13] or lumped parameters [14].

A compromise between computationally intensive numerical methods and equivalent circuits can be achieved using magnetic flux-based models such as those based on the Direct Quadrate (D-Q) model. These models define internal parameters such as flux producing currents and torque producing currents [15]. By modelling internal dynamic factors, a more detailed operational model can be implemented relative to high-level equivalent circuits [16,17], with less computational complexity than numerical method-based techniques.

In this paper, a set of partial differential equations describing a rotating magnetic field and magnetic circuit power transfers will be defined. These equations describe the relationship between magnetic flux and its time derivative. Power transfers due to the magnetic flux-based variables and reluctance are also described. Using these equations and the analytical technique presented in this paper, it is possible to calculate inductive power transfers between rotating objects. These equations can be applied to the analysis of any AC machine incorporating a rotating magnetic field, including variable reluctance configurations.

To demonstrate the validity of this analytical technique, these magnetic circuit equations will be applied to the steady state analysis of an induction motor with results benchmarked against those derived using FEM. It will be based on a more detailed electric motor model than that used to derive the D-Q model. This approach results in a less abstract model that can be derived in terms of closed form equations. While magnetic flux model parameters are usually derived based on dynamic simulations [18,19], the proposed analytical technique parameters will be derived based on motor geometry and the electrical and magnetic material properties.

The proposed analytical technique will only be applied to a specific case in this paper involving steady state analysis. This will, however, provide the basis upon which further applications of the underlying equations can be used to model more complex operational modes, such as those involving transients and faults.

2. Elementary Definitions

To derive a magnetic circuit model of an induction motor, it is necessary to define key parameters to be measured and to describe their behaviour in a environment experiencing circular motion. These techniques are analogous to using circular motion equations to describe rotating objects. Although Newton's second law of motion can be directly applied to a rotating object, using equations of circular motion can considerably reduce the analytical complexity of describing such an object. Similarly, using circular motion magnetic circuit equations can simplify the analysis of rotating electromagnetic devices to the measurement of a few key parameters.

There are four key parameters to be measured when analysing an electromagnetic device incorporating circular motion. Electric current and voltage are two electrical domain parameters with magnetic flux and the time derivative of magnetic flux being their corresponding magnetic domain parameters. In this analysis, electrical domain parameters are mapped onto magnetic domain parameters, then, the analysis is performed in the magnetic domain. To define the relationship between the two magnetic domain parameters, consider the vector representation of a rotating magnetic field as described by (1). This magnetic flux vector is defined using a i, j, k coordinate system, with a radian frequency of ω, a phase offset of θ and time represented by the variable t.

$$\mathbf{\Phi} = |\mathbf{\Phi}|\cos(\omega t + \theta)\hat{\imath} + |\mathbf{\Phi}|\sin(\omega t + \theta)\hat{\jmath} \tag{1}$$

Differentiating (1), with respect to time, allows the derivative of the magnetic flux to be expressed in vector form. For the general case, both the magnitude $|\mathbf{\Phi}|$ and the phase θ will be considered implicit functions of time and the radian frequency ω will be a constant. Using these definitions, the time derivative of (1) can be calculated to be (2).

$$\frac{\partial \mathbf{\Phi}}{\partial t} = \Big(\frac{d|\mathbf{\Phi}|}{dt}\cos(\omega t + \theta) - |\mathbf{\Phi}|(\omega + \frac{d\theta}{dt})\sin(\omega t + \theta)\Big)\hat{\imath} + \Big(\frac{d|\mathbf{\Phi}|}{dt}\sin(\omega t + \theta) + |\mathbf{\Phi}|(\omega + \frac{d\theta}{dt})\cos(\omega t + \theta)\Big)\hat{\jmath} \tag{2}$$

Taking the cross and dot product of the magnetic flux and its time derivative allows (3) and (4), respectively, to be derived.

$$\mathbf{\Phi} \times \frac{\partial \mathbf{\Phi}}{\partial t} = |\mathbf{\Phi}|^2(\omega + \frac{d\theta}{dt}) \tag{3}$$

$$\mathbf{\Phi} \cdot \frac{\partial \mathbf{\Phi}}{\partial t} = |\mathbf{\Phi}|\frac{d|\mathbf{\Phi}|}{dt} \tag{4}$$

It is also possible to use the fundamental properties of the dot and cross product to derive an equation that is independent of the angle between the magnetic flux and its time derivative. This involves squaring both sides of (3) and (4), then adding them together, resulting in (5).

$$\Big|\frac{\partial \mathbf{\Phi}}{\partial t}\Big|^2 = \Big(\frac{d|\mathbf{\Phi}|}{dt}\Big)^2 + |\mathbf{\Phi}|^2(\omega + \frac{d\theta}{dt})^2 \tag{5}$$

Another definition required for this analysis is an angular dependant measurement of resistance. To define this quantity, consider the two conductors coloured grey in Figure 1.

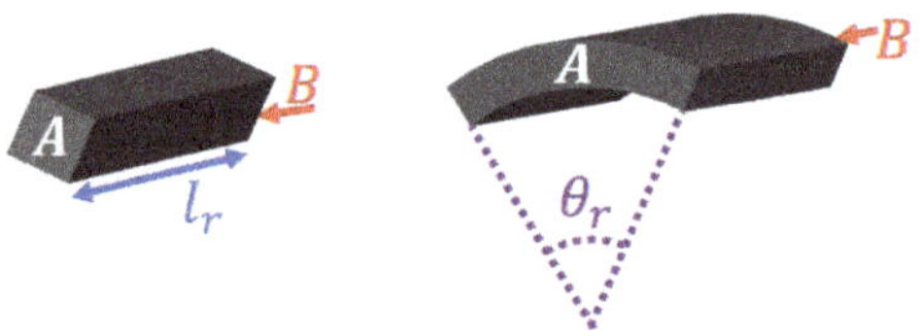

Legend	
Variable	Definition
l_r	Electrical conductor length
θ_r	Angular resistivity measurement angle

Figure 1. Two electrical conductors with dimensions.

In Figure 1, the electrical current flows through each conductor coloured in grey between the surfaces labelled A and B. Assuming the conductive material is homogeneous throughout each volume, the resistance of the left most conductor can be calculated by multiplying its resistance per unit length R_l by its length l_r. Defining the resistance per unit length provides a simple technique to calculate the conductor resistance for an arbitrary length. A similar process can be applied to a change in angular dimension, as demonstrated in the right most conductor. Defining a parameter R_θ with units of Ohm-radians allows the resistance of the right most conductor R between the surfaces labelled A and B to be calculated using (6), where θ_r is defined in Figure 1.

$$R = \frac{R_\theta}{\theta_r} \tag{6}$$

Even if the conductor properties are not homogeneous, R_l can still be used if the conductor is made of a series connection of identical segments, provided the resistance is measured for an integer number of those segments. Similarly, (6) is still valid in situations where periodic angular repetitions occur, provided the resistance is measured for integer multiples of each identical angular segment.

During the operation of an induction motor, energy is transferred from the stator to the rotor via electromagnetic induction. Quantifying the energy transfers due to this process can be achieved using vector-based magnetic circuit equations. The power transfer equation for magnetic circuits [20] is shown in (7).

$$P = \Re\Big(\mathbf{\Phi} \cdot \frac{\partial \mathbf{\Phi}}{\partial t}\Big) + \frac{1}{2}\frac{d\Re}{dt}\Big(\mathbf{\Phi} \cdot \mathbf{\Phi}\Big) \tag{7}$$

For the analysis of an induction motor, magnetic reluctance will be assumed as constant, thereby resulting in the second product term of (7) to equal zero. Therefore, magnetic circuit power transfers can be calculated based on reluctance and both the magnitude and angle of magnetic flux and its time derivative vectors.

Observing magnetic flux vectors from different rotating frames of reference will result in (7) evaluating different values. It has been demonstrated that the induction motor rotor torque can be calculated using the difference in power transfer to the rotor obtained by evaluating (7) from the rotor's and stator's frame of reference [20].

Therefore, frame dependant quantities will be identified in this paper using a right vertical line with a subscript identifying its frame of reference. For example, $\boldsymbol{\Phi}|_{stator}$ and $\boldsymbol{\Phi}|_{rotor}$ are magnetic flux vectors evaluated from the stator's and rotor's frame, respectively. While the magnitude of magnetic flux vectors is frame independent in this analysis, the magnitude and vector value of the magnetic flux time derivative is frame dependant.

For the induction motor derived in this analysis, all rotor electric current-induced magnetic flux will be represented as the single vector $\boldsymbol{\Phi}_r$ and all stator current-induced magnetic flux will be represented by the vector $\boldsymbol{\Phi}_s$. These two vectors are added together using vector addition to form the net magnetic flux $\boldsymbol{\Phi}_n$. Rotor power transfers will then be calculated by substituting the rotor magnetic flux vector into the magnetic flux term in (7) and the time derivative of the net magnetic flux into the flux partial differential term in (7). Repeating this process using stator magnetic flux instead of rotor magnetic flux is used to calculate stator power transfers. A single and constant value of reluctance derived by analysing the magnetic field energy in the motor will be used. Evaluating these power transfers from different rotating frames can then be used to derive an operational model of the induction motor.

3. Rotor and Stator Magnetic Flux Calculations

In an induction motor, magnetic flux is produced by electric currents flowing in the stator windings and inducing currents in the rotor bars. To calculate the rotor's magnetic flux due to induced currents, consider two opposite rotor bars from a cage rotor, as shown in Figure 2.

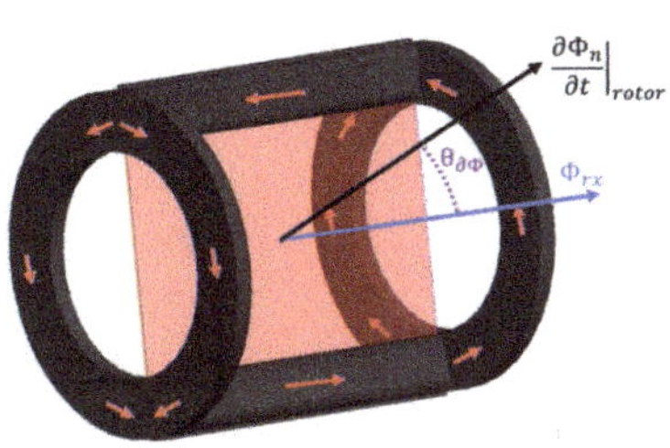

Legend

Variable	Definition
$\left.\frac{\partial \Phi_n}{\partial t}\right\|_{rotor}$	Net Magnetic Flux time derivative evaluated from the rotor frame
$\theta_{\partial\Phi}$	Angular displacement from the partial derivative of the net magnetic flux vector
Φ_{rx}	Dual rotor bar magnetic flux

Figure 2. Two opposite rotor bars with superimposed vectors.

In Figure 2, the electric currents' direction is shown by arrows on the rotor conductors and $\boldsymbol{\Phi}_{rx}$ is the resultant magnetic flux. This magnetic flux will flow through the shaded rectangle with the time derivative of the net magnetic flux from the rotor's frame of reference determining the induced voltage around the conduction loop. This induced voltage V_{rx} can be calculated to be (8), where $\theta_{\partial\Phi}$ is the angular displacement from the partial derivative of the net magnetic flux vector.

$$V_{rx} = \left| \left. \frac{\partial \boldsymbol{\Phi}_n}{\partial t} \right|_{rotor} \right| \cos(\theta_{\partial\Phi}) \tag{8}$$

The induced current in rotor bar pairs can be determined by dividing the induced voltage V_{rx} by the resistance around the conduction loop. The conduction loop resistance, and by extension the rotor's magnetic flux, will depend on the number of rotor bars. When the number of rotor bars is high, this analysis can be simplified using a homogenised model of the rotor bars with the rotor resistance measured in Ohm-radians. To illustrate how this is achieved, consider the cross section of rotor bars as shown in the left diagram in Figure 3.

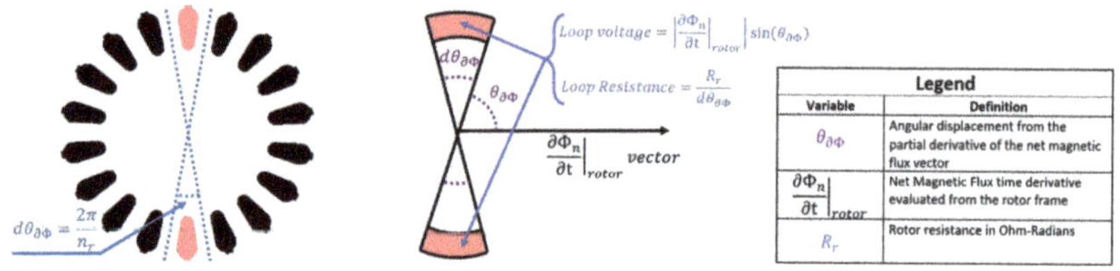

Figure 3. Rotor bar cross sections.

The angular displacement between rotor bars is $\frac{2\pi}{n_r}$, where n_r is the number of rotor bars. As each rotor bar has identical dimensions and angular displacement from its proceeding rotor bar, it is possible to define an angular resistance in Ohm-radians for the rotor R_r by dividing the resistance of the current path as shown in Figure 2 by the angular displacement between rotor bars. This allows the resistance of a conductive loop for a small differential change in angular displacement $d\varphi_{\Phi}$ to be described, as also shown in Figure 3.

Dividing the Figure 3 loop voltage by loop resistance allows the current bounded by the angular displacement of $d\theta_{\partial\Phi}$, as represented by the variable i_{seg}, to be calculated.

$$i_{seg} = \left|\frac{\partial \mathbf{\Phi}_n}{\partial t}\right|_{rotor}\left|\frac{\sin(\theta_{\partial\Phi})}{R_r}d\theta_{\partial\Phi}\right. \quad (9)$$

The total rotor magnetic flux can be determined by combining the contribution of each rotor current segment from Figure 3 into a single magnetic flux vector $\mathbf{\Phi}_r$. This cannot be achieved using a vector addition, as the magnetic field from each rotor segment impacts the other segments. To account for this effect, consider the magnetic field due to two differential pairs of rotor bars and their resultant field directions, as shown in Figure 4.

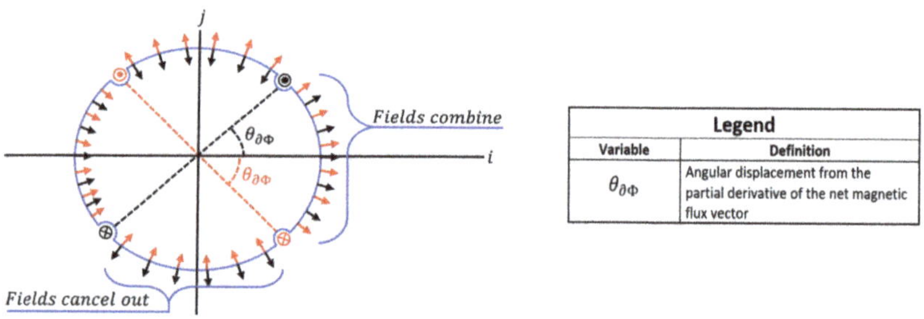

Figure 4. Two differential currents with resultant field directions.

The fields from each current pair will spread approximately evenly around the air gap between the rotor and stator to follow the course of least reluctance. When two current pairs are at the same angular displacement from the *i* axis in Figure 4, components of their fields cancel out while some field components combine. Based on this principle and the geometry from Figure 4, the magnetic flux due to two current pairs $\mathbf{\Phi}_{rxd}$ can be calculated to be (10), where i_{seg} is the current magnitude flowing in each current pair.

$$|\mathbf{\Phi}_{rxd}| = \frac{4 i_{seg}\theta_{\partial\Phi}}{\pi\Re} \quad (10)$$

Assuming the i axis from Figure 4 is aligned with the derivative of the net magnetic flux vector, Equation (10) can be used to calculate the magnetic flux flowing through the rotor using the (9) value of current magnitude.

$$|\mathbf{\Phi}_{rxd}| = \left|\frac{\partial \mathbf{\Phi}_n}{\partial t}\right|_{rotor}\left|\frac{4\theta_{\partial\Phi}\sin(\theta_{\partial\Phi})}{\pi R_r \Re}d\theta_{\partial\Phi}\right. \tag{11}$$

Integrating (11) over the range of $0 \leq \theta_{\partial\Phi} \leq \frac{\pi}{2}$ as shown in Figure 4, results in the (12) value of rotor magnetic flux.

$$|\mathbf{\Phi}_r| = \frac{4}{\pi R_r \Re}\left|\left.\frac{\partial \mathbf{\Phi}_n}{\partial t}\right|_{rotor}\right| \tag{12}$$

The $\mathbf{\Phi}_r$ magnetic flux described by (12) was calculated in the direction of the $\left.\frac{\partial \mathbf{\Phi}_n}{\partial t}\right|_{rotor}$ vector, as shown in the right picture in Figure 3. This is true by definition, as the i axis in Figure 4 is aligned with this vector. It can also be observed in Figure 4 that for any value of $\theta_{\partial\Phi}$, the net magnetic flux along the j axis will be zero. This will result in no component of the rotor magnetic flux vector being perpendicular to the time derivative of the net magnetic flux vector as evaluated in the rotor frame.

As (12) was calculated using the rotor resistance in Ohm-radians, it is independent of the specific number of rotor bars. However, the stator winding configuration cannot be homogenised. Therefore, the relationship between stator currents and induced magnetic flux must be derived separately for every stator winding configuration. For the purpose of demonstrating the model in this paper, a two pole, three phase induction motor stator with 24 rotor slots will be analysed.

Each stator slot contains windings from two phases. For sinusoidal three phase current flowing through the stator windings, the total current amplitude flowing through each stator slot i_s can be calculated to be (13), where i_p is the phase current amplitude and n_s is the number of turns per phase per stator slot.

$$i_s = \sqrt{3}i_p n_s \tag{13}$$

There are 12 different phase slot winding configurations arranged in groups of two. The difference in the current phase between each adjacent group of two is 30°. Based on this information, the (10) formula for magnetic flux can also be applied to stator currents. Substituting (13) into (10) for each differential stator slot pair allows (14) to be derived.

$$|\mathbf{\Phi}_s| = 8.3301\frac{i_p n_s}{\Re} \tag{14}$$

As the number of stator turns per slot per phase n_s and magnetic reluctance $\Re$ are constant in this application, (14) can be simplified using the variable K_i for further analysis, as defined in (15).

$$|\mathbf{\Phi}_s| = K_i i_p \tag{15}$$

4. Modelling Magnetic Reluctance

The conversion of the current into magnetic flux requires the magnetic path reluctance to be known. The magnetic reluctance is calculated using magnetic field energy within the motor as a function of the net magnetic flux $\mathbf{\Phi}_n$. Equation (16) will be used for this calculation, where U is the magnetic field energy.

$$\Re = \frac{2U}{|\mathbf{\Phi}_n|^2} \tag{16}$$

The motor will be divided into 5 regions to calculate magnetic field energy. These concentric regions are shown in the Figure 5 motor cross section, cut perpendicular to the axis of rotation.

In Figure 5, the inner rotor region that falls within the radius of r_i is the section of the rotor that does not contain the rotor bars. This will be modelled with a constant magnetic permeability and parallel magnetic field lines. The total net magnetic flux $\boldsymbol{\Phi}_n$ flows through this region. Due to the assumed parallel nature of the magnetic field lines, the magnetic flux density will be modelled as constant throughout this region. The magnetic flux density within this region can be calculated to be (17), assuming that all of the net magnetic flux $\boldsymbol{\Phi}_n$ passes through this region and the rotor has an axial length of l_m.

$$B_{inner\_rotor} = \frac{|\boldsymbol{\Phi}_n|}{2r_i l_m} \tag{17}$$

Integrating the inner rotor region energy due to the (17) magnetic flux density and substituting this value into (16) allows the (18) value of reluctance to be calculated.

$$\Re_{inner\_rotor} = \frac{\pi}{4\mu_0\mu_r l_m} \tag{18}$$

The rotor bar region and the stator slot region contain both electrical and magnetic field conductors. To make this model easily transferable between different motors, these regions will be modelled homogeneously with each differential angular segment containing an electrical and magnetic conductive region. To achieve this, the ratio of the rotor bar region cross sectional area that contains the rotor bars, relative to its total area, will be defined as C_r.

$$C_r = \frac{rotor\ bar\ area}{rotor\ bar\ region\ area} \tag{19}$$

To illustrate how this concept can be used to calculate the magnetic field energy within the rotor bar region of the rotor, consider the small angular segment of this region bounded by the angular displacement of $d\theta_\Phi$, where θ_Φ is the angular displacement from the net magnetic flux vector.

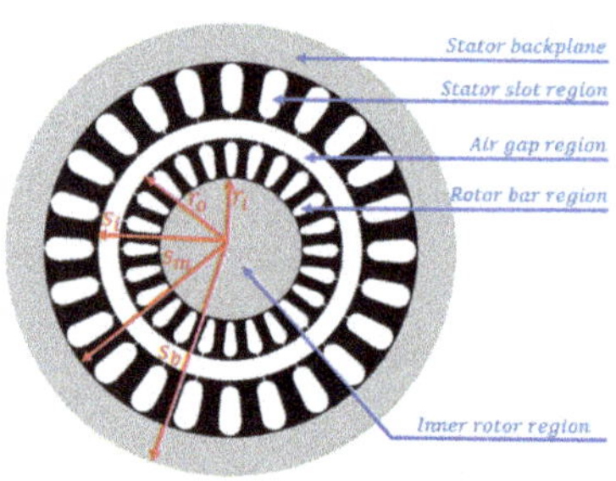

Legend

Variable	Definition
r_i	Inner rotor radius
r_o	Outer rotor radius
s_i	Stator inner radius
s_m	Stator middle radius
s_b	Stator backplane radius

Figure 5. Motor cross section.

In Figure 6, the top shaded region represents the electrical conductor and the lower shaded region represents the magnetic field conductor within this angular segment. Magnetic flux will pass through both the electrical and magnetic conductive regions, although most magnetic flux will pass through the magnetic conductor. The ratio of magnetic flux passing through the electrical to magnetic flux region can be calculated using magnetic circuit equations based on the parallel connection of reluctance elements. Based on this model, the magnetic flux density within the magnetic conductive region B_{mcr} can be calculated as a function of distance from the rotational axis r to be (20).

$$B_{mcr} = \frac{|\boldsymbol{\Phi}_n|\mu_r\cos(\theta_\Phi)}{2l_m r(C_r + \mu_r(1 - C_r))} \tag{20}$$

Integrating this energy over the rotor slot area results in the (21) value of energy within the rotor bar region's magnetic conductors U_{mcr}.

$$U_{mcr} = \frac{\pi \mu_r |\boldsymbol{\Phi_n}|^2 (1 - C_r)}{8\mu_0 l_m (C_r + \mu_r(1 - C_r))^2} log_e\left(\frac{r_o}{r_i}\right) \tag{21}$$

The magnetic flux density in the electrical conductive region B_{ecr} of the rotor bar region can be calculated using the same process to be (22).

$$B_{ecr} = \frac{|\boldsymbol{\Phi_n}| \cos(\theta_\Phi)}{2 l_m r (C_r + \mu_r(1 - C_r))} \tag{22}$$

Integrating this energy over the rotor slot area results in the (23) value of energy within the rotor bar region's electric conductors U_{ecr}.

$$U_{ecr} = \frac{\pi |\boldsymbol{\Phi_n}|^2 C_r}{8\mu_0 l_m (C_r + \mu_r(1 - C_r))^2} log_e\left(\frac{r_o}{r_i}\right) \tag{23}$$

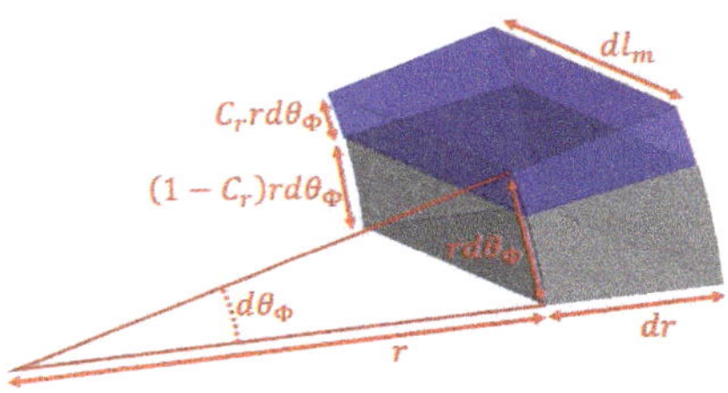

Legend

Variable	Definition
r	Radius variable used for integration
θ_Φ	Angular displacement from net magnetic flux
l_m	Length of rotor and stator
C_r	Rotor bar area to total area ratio

Figure 6. Rotor bar region bound by small angular displacement.

The sum of (21) and (23) is the total magnetic field energy within the rotor bar region of the rotor. Substituting this sum into (16) allows the reluctance of the rotor bar region $\Re_{rotor\_bar}$ to be calculated.

$$\Re_{rotor\_bar} = \frac{\pi}{4\mu_0 l_m (C_r + \mu_r(1 - C_r))} log_e\left(\frac{r_o}{r_i}\right) \tag{24}$$

Calculating the air gap reluctance can be achieved using this same process. Alternatively, substituting $C_r = 1$ into (24) is valid for the air gap region due to its lack of magnetically permeable material. The radius variables r_o and r_i from (24) need to be changed to s_i and r_o, respectively, as per the air gap dimensions in Figure 5, resulting in the (25) value of air gap reluctance.

$$\Re_{air\_gap} = \frac{\pi}{4\mu_0 l_m} log_e\left(\frac{s_i}{r_o}\right) \tag{25}$$

The stator slot region reluctance is calculated using the same technique used to derive (24). This requires the cross sectional area of the stator slots divided by the cross sectional area of the entire stator slot region C_s to be defined.

$$C_s = \frac{\textit{stator slot area}}{\textit{stator slot region area}} \tag{26}$$

Repeating the rotor bar region reluctance calculation using the stator slot area dimensions results in the (27) value of reluctance.

$$\Re_{stator\_slot} = \frac{\pi}{4\mu_0 l_m (C_s + \mu_r(1 - C_s))} log_e\left(\frac{s_m}{s_i}\right) \tag{27}$$

Magnetic flux in the back plane will be the sum of all the magnetic flux that has entered minus what has left. This can be calculated as a function of angular displacement from the net magnetic flux vector θ_Φ by integrating the contribution of each differential flux component leaving the stator slot region.

$$\Phi_{s\_backplane} = \int_{\theta_\Phi}^{0} \frac{|\mathbf{\Phi}_n| dl_m}{2l_m} \cos(\theta_\Phi) d\theta_\Phi \tag{28}$$

$$\Phi_{s\_backplane} = \frac{|\mathbf{\Phi}_n|}{2l_m} \sin(\theta_\Phi) dl_m \tag{29}$$

Dividing (29) by the stator back plane area allows the magnetic flux density in this region to be calculated.

$$B_{s\_backplane} = \frac{|\mathbf{\Phi}_n|}{2l_m(s_o - s_m)} \sin(\theta_\Phi) \tag{30}$$

Integrating the magnetic field energy due to (30) over the entire stator back plane region and substituting the resultant energy value into (16) results in the (31) value of stator back plane magnetic reluctance.

$$\Re_{s\_backplane} = \frac{\pi(s_o + s_m)}{8\mu_0\mu_r(s_o - s_m)l_m} \tag{31}$$

The total equivalent magnetic reluctance $\Re$ of the motor can be calculated by summing together all reluctance elements as described by (32).

$$\Re = \Re_{inner\_rotor} + \Re_{rotor\_bar} + \Re_{air\_gap} + \Re_{stator\_slot} + \Re_{s\_backplane} \tag{32}$$

5. Combining Magnetic and Electric Circuit Power Transfers

Magnetic flux from the rotor and stator will add together to form the net magnetic flux. Due to the sinusoidal distribution of magnetic flux as a function of angular displacement, the stator and rotor magnetic flux vectors can be added using a vector addition.

In a steady state operation of the induction motor, it can be assumed that the net magnetic flux magnitude and the relative angle with its time derivative vector are constant. Therefore, the time derivative of the phase variable θ from (5) can be set to zero. When the derivative of the magnetic flux magnitude is zero, Equation (4) states the angle between the magnetic flux, and its time derivative is 90°. In this situation, Equation (5) can be simplified to equal (33).

$$\left|\frac{\partial \mathbf{\Phi}_n}{\partial t}\right| = |\mathbf{\Phi}_n|\omega \tag{33}$$

The 90° angle between the net magnetic flux and its time derivative will be the same for all rotating frames. However, the net magnetic flux derivative's magnitude is frame dependant due to the ω term in (33).

It was demonstrated in Section 3 that the rotor's induced magnetic flux is parallel to the derivative of the net magnetic flux vector. Using this information, the angular relationship between the stator, rotor and net magnetic flux vectors and the derivative of the net magnetic flux can be depicted in vector form, as shown in Figure 7.

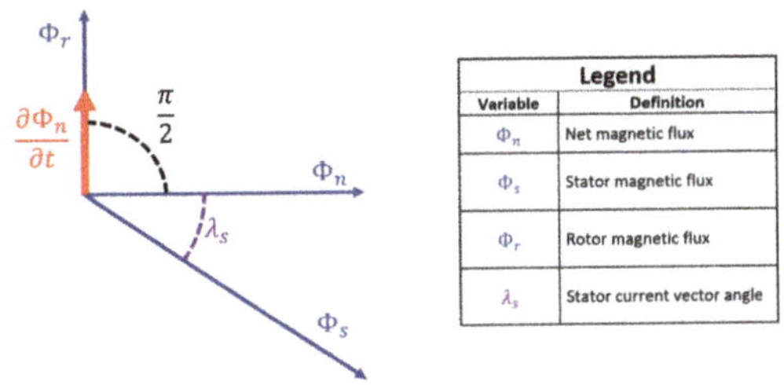

Figure 7. Induction motor magnetic flux vectors.

The stator magnetic flux vector $\mathbf{\Phi_s}$ and the rotor magnetic flux vector $\mathbf{\Phi_r}$ add together to form the net magnetic flux vector $\mathbf{\Phi_n}$. As these vectors represent fields with a sinusoidal air gap distribution relative to angular displacement, they can be summed together using a vector addition. Using the vector geometry from Figure 7, it is possible to derive (34).

$$|\mathbf{\Phi_n}| = |\mathbf{\Phi_s}| \cos(\lambda_s) \tag{34}$$

The magnitude of the net magnetic flux can also be expressed in terms of the phase current amplitude i_p using (15) and (34).

$$|\mathbf{\Phi_n}| = K_i i_p \cos(\lambda_s) \tag{35}$$

Induced voltage is another quantity from the electrical domain that has a magnetic domain equivalent. Just as the variable K_i can be used to calculate stator current from the stator magnetic flux, the variable K_v can be used to calculate the derivative of the net magnetic flux from induced voltage. If the stator winding voltage was only due to the rotating magnetic field, Equation (36) would describe this relationship, where V_s is the stator phase voltage amplitude.

$$\left| \frac{\partial \mathbf{\Phi_n}}{\partial t}^* \right| = K_v V_s \tag{36}$$

However, Equation (36) cannot be exact, as induced voltages exist in the stator windings due to the resistive losses and inductance due to stray magnetic fields. The stator winding voltage due to resistive losses can be calculated by multiplying the stator winding resistance per phase R_s by the stator phase current i_s. As this product is a voltage, it can be converted into the magnetic domain by multiplying by K_v to equal $K_v i_s R_s$. This vector quantity will have a 180° phase shift relative to the stator magnetic flux vector $\mathbf{\Phi_s}$ as per electrical circuit laws. The induced voltage due to the derivative of the net magnetic flux vector, as shown in Figure 7 and described by (33), will have a 180° phase shift as per Lenz's law. Figure 8 shows this information on a vector diagram.

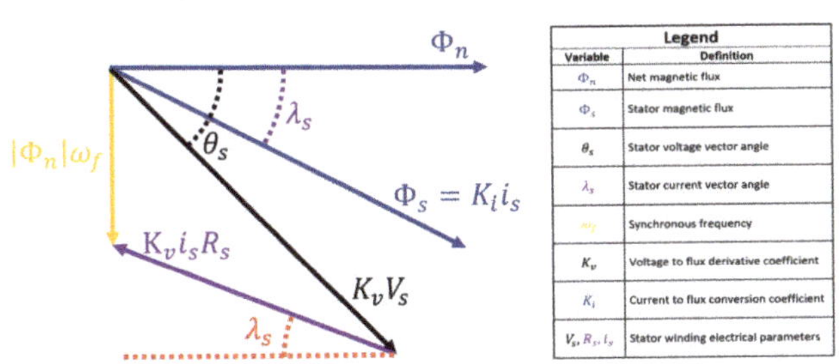

Figure 8. Induction motor magnetic flux and voltage vectors.

In Figure 8, the angle θ_s is the angle between the net magnetic flux and the magnetic flux derivative vector based only on the externally applied stator winding voltage. As this vector diagram is viewed from the stator's frame of reference, the net winding voltage when resistive induced voltages are subtracted from the applied voltage is a function of ω_f, as shown by the left most vector in Figure 8.

Some of the stator windings magnetic flux does not pass through the rotor and therefore cannot be included as part of the stator magnetic flux vector $\mathbf{\Phi}_s$. These stray fields can be modelled as inductors in the series with the stator windings, with an effective inductance of L_S. This effect requires the stator winding voltage drop due to both the resistive losses and stray field inductance V_{R+L} to be accounted for, as described by (37).

$$V_{R+L} = i_s\sqrt{R_S^2 + (\omega_f L_s)^2} \tag{37}$$

The induced winding voltage, as described by (37), will also have a phase shift relative to the purely resistive current induced voltage, as depicted in Figure 8. This additional phase shift is represented by the variable ρ_s, as defined by (38).

$$\rho_s = \tan^{-1}\left(\frac{\omega_f L_s}{R_s}\right) \tag{38}$$

Incorporating this additional information into Figure 8 results in Figure 9.

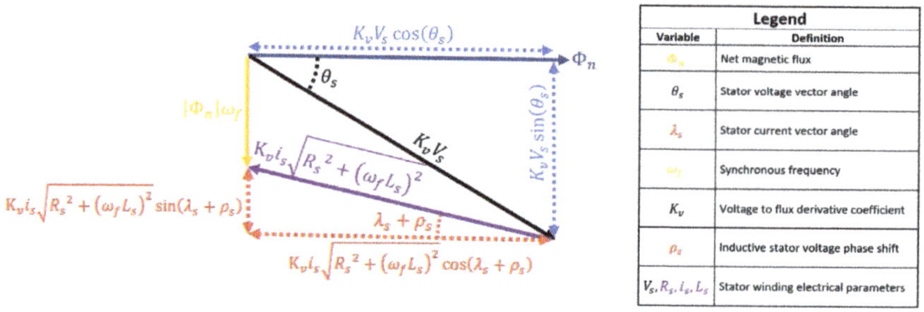

Figure 9. Induction motor voltage vectors.

Equations (39) and (40) can be derived from the Figure 9 vector geometry.

$$V_s \cos(\theta_s) = i_s\sqrt{R_S^2 + (\omega_f L_s)^2}\cos(\lambda_s + \rho_s) \tag{39}$$

$$V_s \sin(\theta_s) = i_s\sqrt{R_S^2 + (\omega_f L_s)^2}\sin(\lambda_s + \rho_s) + \frac{|\mathbf{\Phi}_n|\omega_f}{K_v} \tag{40}$$

The angle between the $K_v V_s$ and $K_i i_s$ vectors from Figure 8 is the same as the phase angle between the AC voltage and current. This is because transferring quantities between the electrical and magnetic domain does not change phases in this example. The AC power delivered to the motor from a three phase source P_{supply} can be calculated to be (41) from elementary electrical identities.

$$P_{supply} = \frac{3}{2}V_s i_s \cos(\theta_s - \lambda_s) \tag{41}$$

Substituting Equations (39) and (40) into (41) results in the (42) value of electrical power.

$$P_{supply} = \frac{3}{2}R_s i_S^2 + \frac{3\omega_f}{2K_v}|\mathbf{\Phi}_n| i_s \sin(\lambda_s) \tag{42}$$

Equation (35) can be substituted into (42) to eliminate i_s in the second product term, resulting in the (43) value of P_{supply}.

$$P_{supply} = \frac{3}{2}R_s i_s^2 + \frac{3\omega_f}{2K_v K_i}|\boldsymbol{\Phi_n}|^2 \tan(\lambda_s) \tag{43}$$

The product of the variables K_i and K_v can be determined using the transforming properties of electrical to magnetic domain conversions. Equation (41) is the power transferred to the stator in the electrical domain. Equating this to its equivalent magnetic domain power transfer equation using (15) and (36) allows the product term $K_i K_v$ to be calculated.

$$\frac{3}{2}V_s i_s \cos(\theta_s - \lambda_s) = \Re\boldsymbol{\Phi_s} \cdot \frac{\partial \boldsymbol{\Phi_n}}{\partial t}^* \tag{44}$$

$$K_i K_v = \frac{3}{2\Re} \tag{45}$$

Substituting (45) into (43) results in the (46) value of stator electrical power.

$$P_{supply} = \frac{3}{2}R_s i_s^2 + \Re\omega_f|\boldsymbol{\Phi_n}|^2 \tan(\lambda_s) \tag{46}$$

6. Power Losses

Power supplied to the induction motor can be described by (46), although at this stage in the derivation, variables λ_s, i_s and $\boldsymbol{\Phi_n}$ are unknown. To calculate these variables, all sources of internal losses need to be quantified. Magnetic losses resulting from eddy currents and hysteresis losses can be quantified using the Steinmetz's equation. For this model, stator magnetic power losses P_{sm} will be approximated using (47), while rotor magnetic losses P_{rm} will be approximated using (48).

$$P_{sm} = M_{se}\omega_f^2|\boldsymbol{\Phi_n}|^2 + M_{sh}\omega_f|\boldsymbol{\Phi_n}|^2 \tag{47}$$

$$P_{rm} = M_{re}\omega_s^2|\boldsymbol{\Phi_n}|^2 + M_{rh}\omega_s|\boldsymbol{\Phi_n}|^2 \tag{48}$$

In Equations (47) and (48), the variables M_{se} and M_{re} are the coefficients of stator eddy current losses and rotor eddy current losses respectively. M_{sh} and M_{rh} are the coefficients of stator hysteresis losses and rotor hysteresis losses, respectively.

Rotor electrical power losses can be calculated using the magnetic circuit power transfer equation, as described by (7). Substituting the (12) value of the rotor magnetic flux and the (33) derivative of the magnetic flux evaluated from the rotor's reference frame into (7) results in the (49) value of the rotor electric power loss P_{re}.

$$P_{re}|_{rotor} = \frac{4}{\pi R_r}\omega_s^2|\boldsymbol{\Phi_n}|^2 \tag{49}$$

Evaluating P_{re} from the stator's reference frame results in the (50) value of power loss.

$$P_{re}|_{stator} = \frac{4}{\pi R_r}\omega_s\omega_f|\boldsymbol{\Phi_n}|^2 \tag{50}$$

The difference between the rotor electric power loss when evaluated from the stator's and rotor's frame of reference is the rate electrical energy is converted into mechanical energy due to the rotor torque. Only electrical losses need to be accounted for in this calculation, as they produce the mutually coupled magnetic flux between the rotor and stator.

Based on this information, the power transfer due to torque on the rotor can be calculated to be (51), where ω_r is the rotor rotational speed $(\omega_f - \omega_s)$.

$$P_\tau = \frac{4}{\pi R_r}\omega_s\omega_r|\boldsymbol{\Phi_n}|^2 \tag{51}$$

7. Complete Power Transfer Equations

Equation (46) quantifies the rate electrical energy is transferred to the motor and is made up of two product terms. The first product term describes energy lost due to the electrical resistance of the stator windings, while the second describes inductive energy transfers. These inductive energy transfers are the (50) rotor power and the (47) and (48) magnetic losses. Equating the second product term of (46) to equal these inductive energy transfers allows the value of $\tan(\lambda_s)$ to be calculated.

$$\tan(\lambda_s) = \frac{4\omega_s}{\pi R_r \Re} + \frac{M_{se}\omega_f^2 + M_{sh}\omega_f + M_{re}\omega_s^2 + M_{rh}\omega_s}{\omega_f \Re} \tag{52}$$

The value of λ_s can now be calculated using (52), as it is a function of known constants and externally measurable variables.

It is also necessary to calculate the magnitude of the net magnetic flux $\mathbf{\Phi}_n$. This can be achieved by first squaring (39) and (40), then adding them together to eliminate their dependence on θ_s.

$$V_s^2 = i_s^2\left(R_s^2 + (\omega_f L_s)^2\right) + \left(\frac{\omega_f|\mathbf{\Phi}_n|}{K_v}\right)^2 + \frac{2 i_s \omega_f}{K_v}\sin(\lambda_s + \rho_s)\sqrt{R_s^2 + (\omega_f L_s)^2} \tag{53}$$

Substituting Equations (35), (38) and (45) into Equation (53) allows the i_s, ρ_s and K_v terms to be eliminated, resulting in (54).

$$\frac{V_s^2}{|\mathbf{\Phi}_n|^2} = \frac{R_s^2 + (\omega_f L_s)^2}{K_i^2}\left(\tan^2(\lambda_s) + 1\right) + \left(\frac{2\Re K_i \omega_f}{3}\right)^2 + \frac{4\Re\omega_f}{3}\left(R_s \tan(\lambda_s) + \omega_f L_s\right) \tag{54}$$

As the amplitude of the stator winding voltage V_s is known, Equation (54) can be used to calculate $|\mathbf{\Phi}_n|$. This value can then be substituted into (51) to calculate the power transfer due to the rotor torque. Electrical power supplied to the motor can be calculated using (46) and the stator phase current using (35).

8. FEM Simulation Parameters

The equations derived in this paper can be used to predict the torque speed characteristics of an induction motor. To validate the model derived in this paper, a FEM simulation of an induction motor's torque speed characteristics will be used to establish a benchmark against which the accuracy of these equations can be assessed. The following geometric parameters as described in Section 4 of this paper used in the FEM simulation are as follows.

- $r_i = 43$ mm
- $r_o = 65$ mm
- $s_i = 65.5$ mm
- $s_m = 87$ mm
- $s_b = 112$ mm
- $l_m = 160$ mm
- $c_r = 0.65$
- $c_s = 0.65$
- $\mu_r = 3000$

The relative permeability of non-oriented superior motor grade electrical steel is used in the simulation due to its popularity in small- to mid-sized motors [21]. Soft magnetic materials such as silicon steel (3.2 wt% Si) are increasing in popularity due to low loss factors. However, high processing costs and the difficulty in optimising all required mechanical and electromagnetic parameters is still a limiting factor in the commercialisation of soft magnetic materials [22]. As this analysis does not require a complex magnetic model to demonstrate its validity, the magnetic properties of conventional motor grade electrical steel will be used where required.

Inserting these dimensions into the Section 4 magnetic reluctance equations results in the following reluctance values.

- $\Re_{inner\_rotor} = 1302\ \text{H}^{-1}$
- $\Re_{rotor\_bar} = 1536\ \text{H}^{-1}$
- $\Re_{air\_gap} = 29{,}933\ \text{H}^{-1}$
- $\Re_{stator\_slot} = 1055\ \text{H}^{-1}$
- $\Re_{s\_backplane} = 5182\ \text{H}^{-1}$

The sum of these motor reluctance values results in $\Re = 39{,}008\ \text{H}^{-1}$.

Not all the magnetic flux produced by stator currents is mutually coupled with the rotor. This results in stator leakage flux. Quantifying this value is required to calculate the stator windings leakage inductance. Leakage flux can be quantified using FEM [23] or real time emulation models [24]. Magnetic circuit-based leakage flux models are also effective and can be derived based on geometric factors [25]. As such, the leakage inductance will be calculated in this model using a reluctance mesh.

This value is determined by estimating the magnetic path reluctance encountered by these stray fields. The path encountered by one set of stray fields is shown in Figure 10 by the reluctance circuit surrounding a stator slot.

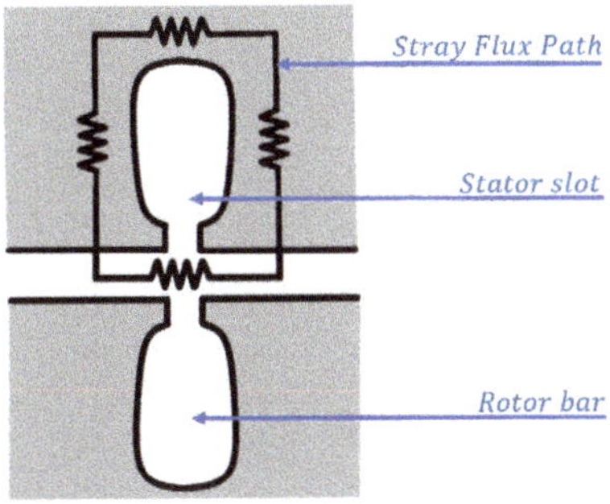

Figure 10. Stray fields' magnetic flux path.

Figure 10 depicts the reluctance circuit surrounding a single stator slot. As a phase winding current will flow through four adjacent slots, the parallel connection of four reluctance circuits as depicted by Figure 10 will need to be calculated. Using the resulting reluctance mesh, the reluctance encountered by stray flux is estimated to be $2.85 \times 10^6\ \text{H}^{-1}$. The induction motor is wound with 26 windings per phase per stator slot. This results in 104 conductors enclosed by the stray fields' reluctance mesh. Therefore, the effective series inductance in the stator windings due to leakage flux can be calculated using (55).

$$L = \frac{n^2}{\Re} \tag{55}$$

In (55), L is inductance, n is the number of turns and $\Re$ is reluctance. Using this relationship, the inductance due to a single phase winding passing through four adjacent stator slots is 3.80 mH. As each phase has four sets of these windings, the leakage inductance per stator phase winding, as represented by the variable L_s, is 15.2 mH.

The rotor resistance can be calculated by measuring the electrical path resistance of a current loop encompassing two opposite rotor bars, as shown in Figure 2. This resistance for the simulated rotor is estimated to be $1.14 \times 10^{-4}\ \Omega$. Multiplying this value by the angular displacement between rotor bars $(\frac{\pi}{12})$ results in the rotor resistance R_r value of 2.98×10^{-5} Ohm-radians. The stator winding resistance R_s has been set in the simulation to be 1.616 Ω. Other simulation parameters are a phase to phase RMS voltage of 400 V, a Wye connection and a synchronous frequency of 100π radians per second.

For this simulation, all magnetic loss coefficients will be set to zero. This is to validate the model derivation up to this point. Deriving magnetic loss coefficients directly from motor geometry and magnetic material properties will require an analysis of approximately the same length as this paper up to this point. Alternatively, using FEM to derive approximations for magnetic loss parameters would be using FEM to validate its own modelling technique. This approach would also invalidate the claim in the introduction that all parameters can be derived directly from physical laws without the aid of existing simulation models.

9. FEM Simulation Results

Predictions made by the magnetic circuit-based model presented in this article will be benchmarked against a commercially available FEM model to assess its accuracy. This is achieved using the FEM simulation tools of Ansys Maxwell 16.0. A 3D rendering of the simulated motor model is shown in Figure 11 from two angles with phase windings colour-coded black, red and blue for phases 1, 2 and 3, respectively. The motor model is cut both radially and axially, as the entire motor geometry is not required to be simulated due to symmetries.

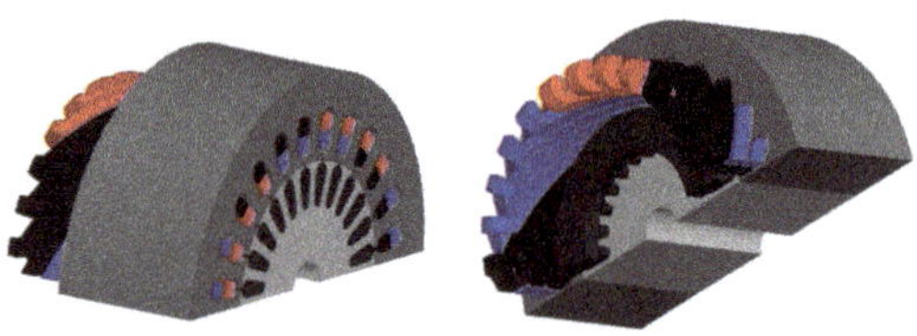

Figure 11. Three-dimensional rendering of FEM-simulated induction motor.

Using the parameters detailed in the previous section, a FEM simulation of the induction motor was undertaken to obtain its output power, phase current and efficiency as a function of rotor rotational speed. A 2D rendering of the FEM simulated motor's magnetic fields is shown in Figure 12 once they have reached their steady state values.

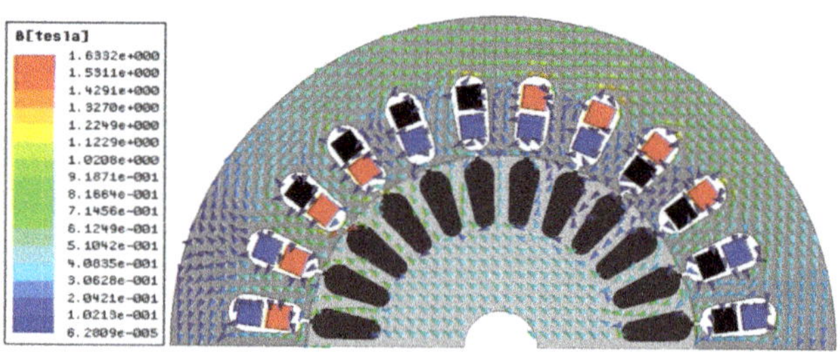

Figure 12. Two-dimensional rendering of FEM-simulated induction motor with superimposed fields.

This rendering validates the assumptions of the magnetic field distribution within the motor, as stated in Section 4. Magnetic field distribution within the inner rotor region is approximately uniform in both magnitude and direction. Stator fields also vary in intensity, reaching their peak values perpendicular to the direction of the inner rotor's magnetic field from the axis of rotation.

The FEM simulation also allowed the output power as a function of rotational speed to be calculated, as shown in Figure 13. These simulated values are compared to those obtained using the magnetic circuit model derived in this paper, with the rotational speed measured in Revolutions Per Minute (RPM).

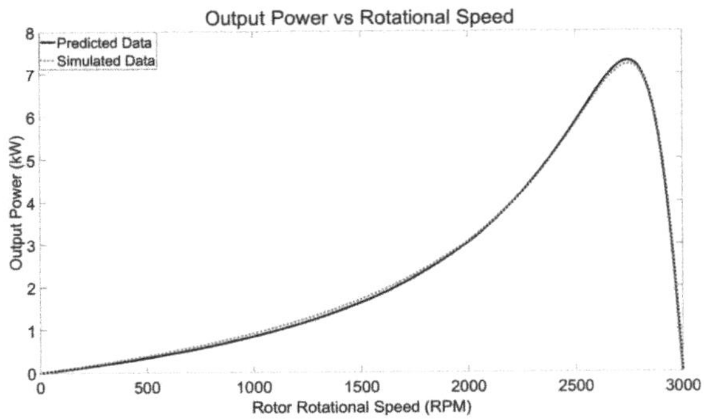

Figure 13. Output power vs. rotor rotational speed.

The simulated output power in Figure 13 slightly exceeds the predicted value for rotational speeds between 500 RPM and 2000 RPM. However, the maximum predicted output power was 7.29 kW, which is 70 W more than the simulated value of 7.22 kW. For rotational speeds above 2800 RPM, which correspond to typical operating conditions, the FEM-simulated and magnetic circuit predictions are closely matched. The RMS deviation between the two output power data sets is 59.3 W or 2.47% of the average predicted value.

As the magnetic circuit parameters derived in this model do not correspond directly with those used in the FEM simulation, it is difficult to identify which parameter caused this discrepancy. It is worth noting that although the FEM-simulated magnetic losses were set to zero in the material settings, a low but non zero value of magnetic loss was recorded. This implies that the FEM model uses magnetic loss correction techniques to improve its accuracy in practical settings.

The predicted phase current in Figure 14 is less than the FEM-simulated values for rotational speeds below 1500 RPM. Both data sets are closely matched for rotational speeds above 1500 RPM with only a slight divergence around 2400 RPM. Overall, the RMS deviation between the two data sets is 0.223 A or 0.87% of the average predicted current.

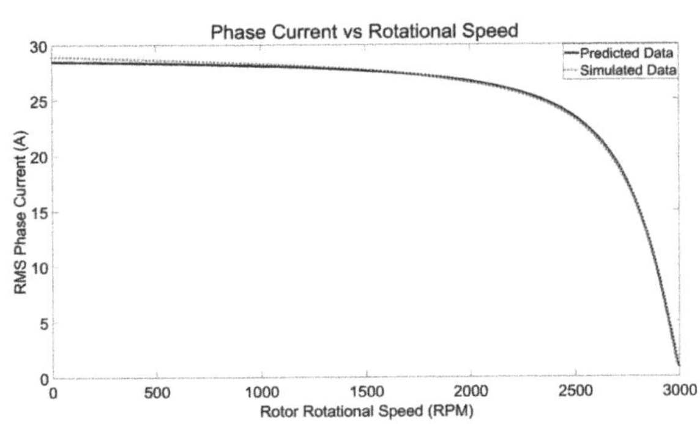

Figure 14. Phase current vs. rotor rotational speed.

The predicted motor efficiency in Figure 15 was slightly less than its simulated value for most rotational speeds. However, the peak predicted efficiency of 95.8% exceeds the maximum simulated efficiency of 94.3%. Overall, the RMS deviation between the two data sets is 0.92%.

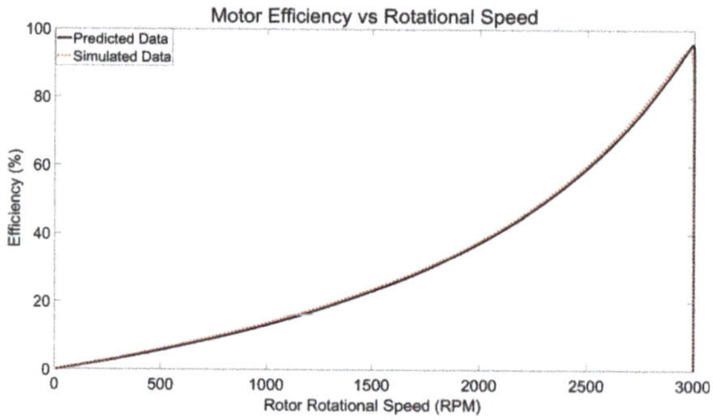

Figure 15. Motor efficiency vs. rotor rotational speed.

Figure 15 also reveals another possible reason for the differences between the FEM and magnetic circuit-derived results. The FEM efficiency is lower than magnetic circuit efficiency close to the synchronous frequency. In this operational region, the stator losses dominate; thus, FEM stator losses might be higher than the magnetic circuit model. As overall FEM efficiency is slightly higher, a larger FEM inductive coupling factor or lower FEM rotor losses may also be responsible.

The efficiency function in Figure 15 increases with rotational speed across most of the data set range, before quickly reducing to zero at the synchronous speed. This overall shape is to be expected and can be inferred from the rotor power transfer Equations (50) and (51). Only factoring these into the efficiency equation allows an upper limit on motor efficiency to be derived. This upper limit is (56), where η is the motor efficiency.

$$\eta \leq 1 - \frac{\omega_s}{\omega_f} \quad (56)$$

In (56), ω_s is the slip frequency, which is the synchronous frequency minus the rotational frequency. Equation (56) represents the theoretical limit of the inductive transfer efficiency to the rotor with a constant reluctance. To achieve this efficiency, no stator losses can be incurred. As such, the deviation in efficiency from (56) can reveal the impact of stator related and mechanical sources of power loss on overall motor efficiency.

It is also possible to calculate the angle between the net and stator magnetic flux vectors λ_s as a function of rotor rotational speed. As this is not a variable commonly measured by simulation software, plotting this information offers a unique insight into the internal magnetic field dynamics within the induction motor. The angle λ_s is an important parameter, as it describes how much stator magnetic flux, and therefore, the stator winding current, is needed to produce the net magnetic flux, as per (34). It also determines how much stator winding current is required to achieve inductive power transfers. This value can be determined by substituting (34) into (46).

The value of λ_s is shown in Figure 16 as a function of rotor rotational velocity.

It can be observed from Figure 16 that the angle between the stator and net magnetic flux vectors λ_s is close to 90° for low rotor speeds and quickly approaches 0° near the synchronous rotational speed. It is also interesting to note that the operating speeds most likely to be encountered under normal load have the most variation in λ_s.

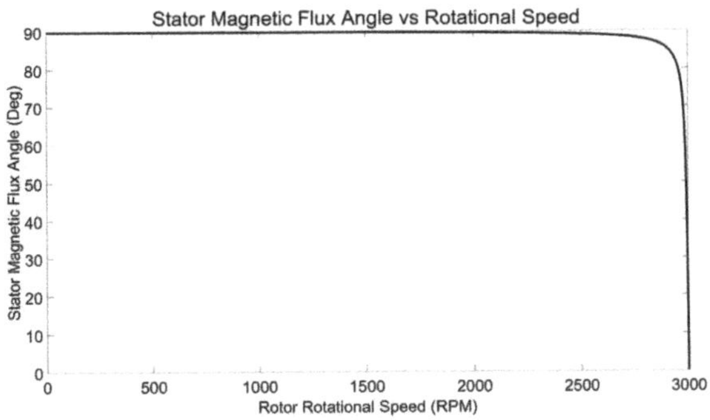

Figure 16. Angle between net and stator magnetic flux vectors vs. rotor rotational speed.

Figure 16 also reveals that the relative angles between magnetic flux vectors can change with rotational speed. Other magnetic flux models such as the D-Q model have a fixed angle of 90° between their two magnetic flux vectors. As the angle between the rotor and net magnetic flux vectors is 90°, it is possible to see why this assumption might still result in accurate observations. However, as per (4), this constant 90° angle is only possible when the derivative of the net magnetic flux magnitude is zero, as the rotor magnetic flux is parallel to the time derivative of the net magnetic flux. Any change to net magnetic flux magnitude will also change this angle, irrespective of the angular frequency ω and phase θ, as (4) is independent of these variables. Accounting for this change in angle between magnetic flux vectors could therefore provide this model with an advantage over fixed angle models.

The net magnetic flux magnitude $|\boldsymbol{\Phi}_n|$ can be readily obtained from the magnetic circuit equations derived in this paper. This value is shown in Figure 17 as a function of rotor rotational speed.

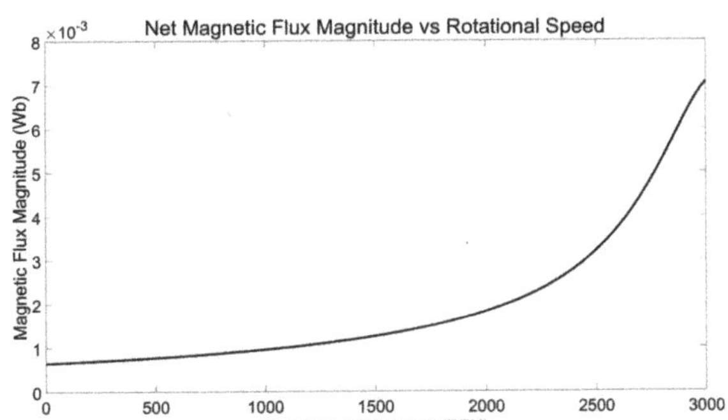

Figure 17. Net magnetic flux magnitude vs. rotor rotational speed.

Figure 17 depicts a continual increase in the net magnetic flux magnitude as a function of rotor rotational speed. The stator magnetic flux, which is directly proportional to the stator current, as shown in Figure 14, only decreases when the rotational speed increases. This illustrates the impact the angle between the net and stator magnetic flux has on the operation of the induction motor. For low speeds, the stator and rotor magnetic flux are almost opposite each other and therefore much of their magnetic flux is canceled out. As the angle λ_s decreases, less stator flux and therefore phase current is required to produce the net magnetic flux.

10. Conclusions

This paper has demonstrated how magnetic circuit analysis can be applied to situations involving rotating magnetic fields using energy and circular motion-based equations. The key components of this analytical technique involves both magnetic flux and the time derivative of magnetic flux in vector form. Quantifying these magnetic domain quantities from their electrical domain equivalents and using the mathematical relationship between them allows for the derivation of closed form equations. The resultant magnetic circuit model can then be used to describe the motor's operational characteristics and provide insights into parameters not usually quantified in traditional models.

As a magnetic flux-based model, the proposed analytical technique bears some similarities to the D-Q model. One advantage of this model over the D-Q model is that parameters can be derived directly from motor geometry. Another is that it is derived at a lower level of abstraction than the assumptions upon which the D-Q model is based. As such, it involves a more fundamental approach, from the application of the laws of physics to the description of AC machines.

It has also been demonstrated in this paper that magnetic circuit analysis can be used to describe rotating machines without the need for ad hoc or unexplained variables to match predictions with observations. The same parameters and variables were used to derive all of the predicted data without the need for individual correction coefficients for each measurements.

Due to the closed form equations derived using this analytical technique, there is an inherent computational speed advantage of this model over numerical method-based models. This could be advantageous for future applications of the underlying mathematical relationships described in this paper if implemented in real time control applications. It can also provide motor designers with information such as relative magnetic flux vector angles, which cannot be readily obtained from some design software.

The model presented in this paper only represents one specific application of these equations relating the magnetic flux and its time derivative. Therefore, there are still further applications of this research using the underlying equations and analytical techniques to derive dynamic models incorporating more complex losses. There is also the potential to use these magnetic circuit equations as a basis on which to build control algorithms, which has not yet been explored. Through further applications of the equations derived in this paper, it may be possible to use magnetic circuit analysis to achieve a more optimal balance between computational efficiency and accuracy in practical applications.

Funding: This research was funded through the provision of an Australian Government Research Training Program Scholarship, a Monash University Faculty of Information Technology Research Scholarship and a CSIRO Data61 Scholarship.

Conflicts of Interest: The author declares no conflict of interest. The funders had no role in the design of the study; in the collection, analyses, or interpretation of data; in the writing of the manuscript, or in the decision to publish the results.

Abbreviations

Symbol	Quantity
B	Magnetic flux density
B_{inner_rotor}	Inner rotor magnetic flux density
$B_{s\_backplane}$	Stator backplane magnetic flux density
B_{ecr}	B field in rotor bar area electric conductor
B_{mcr}	B field in rotor bar area magnetic conductor
C_r	Rotor bar area to total area ratio
C_s	Stator slot area to total area ratio

i_p	Phase current amplitude
i_s	Stator slot current amplitude
i_{seg}	Current in differential rotor segment
K_i	Current to flux conversion coefficient
K_v	Voltage to flux derivative coefficient
L	Inductance value
l_m	Length of rotor and stator
l_r	Electrical conductor length
L_s	Stator winding inductance
M_{re}	Coefficient of rotor eddy current losses
M_{rh}	Coefficient of rotor hysteresis losses
M_{se}	Coefficient of stator eddy current losses
M_{sh}	Coefficient of stator hysteresis losses
n	Number of winding turns
n_s	Winding turns per stator slot per phase
P	Generic power variable
P_{re}	Rotor electrical power loss
P_{rm}	Rotor magnetic eddy current losses
P_{rotor}	Power transfer to the rotor
P_{sm}	Stator magnetic eddy current losses
P_{stator}	Power transfer from stator
P_{supply}	Electrical power supplied to motor
P_{τ}	Mechanical power transfer due to rotor torque
r	Radius variable used for integration
r_i	Inner rotor radius
r_o	Outer rotor radius
R	Generic resistance value
R_l	Linear resistivity
R_r	Rotor angular resistance
R_s	Stator winding resistance
R_{θ}	Angular resistivity
$\Re$	Magnetic reluctance
$\Re_{air\_gap}$	Air gap reluctance
$\Re_{inner\_rotor}$	Inner rotor reluctance
$\Re_{rotor\_bar}$	Rotor bar area reluctance
$\Re_{s\_backplane}$	Stator backplane reluctance
$\Re_{stator\_slot}$	Stator slot reluctance
S_b	Stator backplane radius
S_i	Stator inner radius
S_m	Stator middle radius
t	Time
U	Generic energy variable
U_{ecr}	Energy in rotor bar area electric conductor
U_{mcr}	Energy in rotor bar area magnetic conductor
V_{rx}	Induced voltage in rotor bar pair x
V_{R+L}	Stator resistive and inductive voltage
V_s	Stator voltage amplitude
η	Rotor power transfer efficiency
θ	Generic angle variable
θ_r	Angular resistivity measurement angle
$\theta_{\partial\Phi}$	Angular displacement from $\mathbf{\Phi}_n$ partial derivative
θ_{Φ}	Angular displacement from $\mathbf{\Phi}_n$
θ_s	Stator voltage vector angle
λ_s	Stator current vector angle
μ_r	Relative permeability
μ_0	Vacuum permeability
ρ_s	Inductive stator voltage phase shift

$\boldsymbol{\Phi}$	Generic magnetic flux variable
$\boldsymbol{\Phi}_n$	Net magnetic flux
$\boldsymbol{\Phi}_s$	Stator magnetic flux
$\boldsymbol{\Phi}_r$	Rotor magnetic flux
$\boldsymbol{\Phi}_{rx}$	Dual rotor bar magnetic flux
$\boldsymbol{\Phi}_{rxd}$	Differential rotor magnetic flux
$\Phi_{s\_backplane}$	Stator backplane magnetic flux
ω	Generic frequency variable
ω_f	Synchronous frequency
ω_s	Slip frequency
ω_r	Rotor rotational frequency

References

1. Fan, M.; Chai, J.; Sun, X. Induction motor parameter identification based on T-model equivalent circuit. In Proceedings of the 17th International Conference on Electrical Machines and Systems ICEMS, Hangzhou, China, 22–25 October 2014; pp. 2535–2539.
2. Bhowmick, D.; Manna, M.; Chowdhury, S.K. Improved equivalent circuit parameter estimation of induction motor using HG diagram and PSO. In Proceedings of the IEEE Calcutta Conference CALCON, Kolkata, India, 2–3 December 2017; pp. 443–447.
3. Ling, Z.; Zhou, L.; Guo, S.; Zhang, Y. Equivalent circuit parameters calculation of induction motor by finite element analysis. *IEEE Trans. Magn.* **2014**, *50*, 833–836. [CrossRef]
4. Wang, Y.J.; Lee, M.H. Analytical modeling of the pulsation torque of an induction motor under Steinmetz circuit. In Proceedings of the 3rd IEEE Conference on Industrial Electronics and Applications, Singapore, 3–5 June 2008; pp. 1464–1468.
5. Dalal, A.; Singh, A.K.; Kumar, P. Effect of saturation on equivalent circuit analysis of induction motor in practical scenario. In Proceedings of the 2013 Annual IEEE India Conference INDICON, Mumbai, India, 13–15 December 2013; pp. 1–5.
6. Mohammadi, A.S.; Trovão, J.P. A comparison of different models for permanent magnet synchronous machines: Finite element analysis, DQ lumped parameter modeling, and magnetic equivalent circuit. In Proceedings of the IEEE 28th International Symposium on Industrial Electronics ISIE, Vancouver, BC, Canada, 12–14 June 2019; pp. 197–202.
7. Chen, Q.; Liu, G.; Zhao, W.; Shao, M. Nonlinear adaptive lumped parameter magnetic circuit analysis for spoke-type fault-tolerant permanent-magnet motors. *IEEE Trans. Magn.* **2013**, *49*, 5150–5157. [CrossRef]
8. Han, Q.; Wang, T.; Ding, Z.; Xu, X.; Chu, F. Magnetic equivalent modeling of stator currents for localized fault detection of planetary gearboxes coupled to electric motors. *IEEE Trans. Ind. Electron.* **2020**, *68*, 2575–2586. [CrossRef]
9. Bouzida, A.; Abdelli, R.; Boudouda, A.; Bendjghaba, O.; Ouadah, M.H. Lumped-Parameter Thermal Model of Induction Motor under Phase Open-Circuit Fault. In Proceedings of the 2019 International Conference on Advanced Electrical Engineering ICAEE, Algiers, Algeria, 19–21 November 2019; pp. 1–6.
10. Amrhein, M.; Krein, P.T. Induction machine modeling approach based on 3-D magnetic equivalent circuit framework. *IEEE Trans. Energy Convers.* **2010**, *25*, 339–347. [CrossRef]
11. Faiz, J.; Ebrahimi, B.M.; Akin, B.; Toliyat, H.A. Finite-element transient analysis of induction motors under mixed eccentricity fault. *IEEE Trans. Magn.* **2007**, *44*, 66–74. [CrossRef]
12. Sudhoff, S.D.; Kuhn, B.T.; Corzine, K.A.; Branecky, B.T. Magnetic equivalent circuit modeling of induction motors. *IEEE Trans. Energy Convers.* **2007**, *22*, 259–270. [CrossRef]
13. Amiri, E. Study of linear induction motor end effects using 3-D FEM and equivalent circuit. In Proceedings of the 39th Annual Conference of the IEEE Industrial Electronics Society IECON, Vienna, Austria, 10–13 November 2013; pp. 3165–3170.
14. Di Leonardo, L.; Popescu, M.; Tursini, M.; Villani, M. Finite elements model co-simulation of an induction motor drive for traction application. In Proceedings of the 45th Annual Conference of the IEEE Industrial Electronics Society IECON, Lisbon, Portugal, 14–17 October 2019; pp. 1059–1065.
15. Popescu, M. *Induction Motor Modelling for Vector Control Purposes*; Helsinki University of Technology: Espoo, Finland, 2000.
16. El-Faouri, F.S.; Mohamed, O.; Elhaija, W.A. DQ model and control of a three-phase induction motor considering mutual flux saturation effect. In Proceedings of the 10th Jordanian International Electrical and Electronics Engineering Conference JIEEEC, Amman, Jordan, 16–17 May 2017; pp. 1–6.
17. Hanif, A.; Ali, S.N.; Ahmed, Q.; Bhatti, A.I.; Yin, G.; Jaffery, M.H. Effect of variation in rotor resistance on the dynamic performance of induction motor. In Proceedings of the 35th Chinese Control Conference CCC, Chengdu, China, 27–29 July 2016; pp. 9524–9529.
18. Usama, M.; Kim, J. Vector control algorithm based on different current control switching techniques for Ac motor drives. *arXiv* **2020**, arXiv:2005.04651.
19. Alawan, M.A.; Al-Subeeh, A.N.N.; Al-Furaiji, O.J.M. Simulating an induction motor multi-operating point speed control using PI controller with neural network. *Period. Eng. Nat. Sci.* **2019**, *7*, 1478–1485.
20. Kidd, B. Modelling of Inhomogeneous Magnetic Fields in Electromagnetic Devices. Master's Thesis, RMIT University, Melbourne, Australia, 2018.
21. Beckley, P.; Sujan, G.K. *Steels, Silicon Iron-Based: Magnetic Properties, Reference Module in Materials Science and Materials Engineering*; Elsevier: Amsterdam, The Netherlands, 2016; pp. 1–5.

22. Ouyang, G.; Chen, X.; Liang, Y.; Macziewski, C.; Cui, J. Review of Fe-6.5 wt%Si high silicon steel—A promising soft magnetic material for sub-kHz application. *J. Magn. Magn. Mater.* **2019**, *481*, 234–250. [CrossRef]
23. Cheaytani, J.; Benabou, A.; Tounzi, A.; Dessoude, M.; Chevallier, L.; Henneron, T. End-Region Leakage Fluxes and Losses Analysis of Cage Induction Motors Using 3-D Finite-Element Method. *IEEE Trans. Magn.* **2015**, *51*, 1–4. [CrossRef]
24. Masadeh, M.A.; Pillay, P. Induction motor emulation including main and leakage flux saturation effects. In Proceedings of the IEEE International Electric Machines and Drives Conference IEMDC, Miami, FL, USA, 21–24 May 2017; pp. 1–7.
25. Nishihama, K.; Ide, K.; Mikami, H.; Fujigaki, T.; Mizutani, S. Starting torque analysis of cage induction motor using permeance model considering magnetic saturation by leakage flux. In Proceedings of the International Conference on Electrical Machines and Systems, Tokyo, Japan, 15–18 November 2009; pp. 1–6.

MDPI AG
Grosspeteranlage 5
4052 Basel
Switzerland
Tel.: +41 61 683 77 34

Magnetism Editorial Office
E-mail: magnetism@mdpi.com
www.mdpi.com/journal/magnetism

www.ingramcontent.com/pod-product-compliance
Lightning Source LLC
Chambersburg PA
CBHW061626210926
51850CB00009B/2456
9783725870745